CHEMICAL OCEANOGRAPHY

CHEMICAL
OCEANOGRAPHY

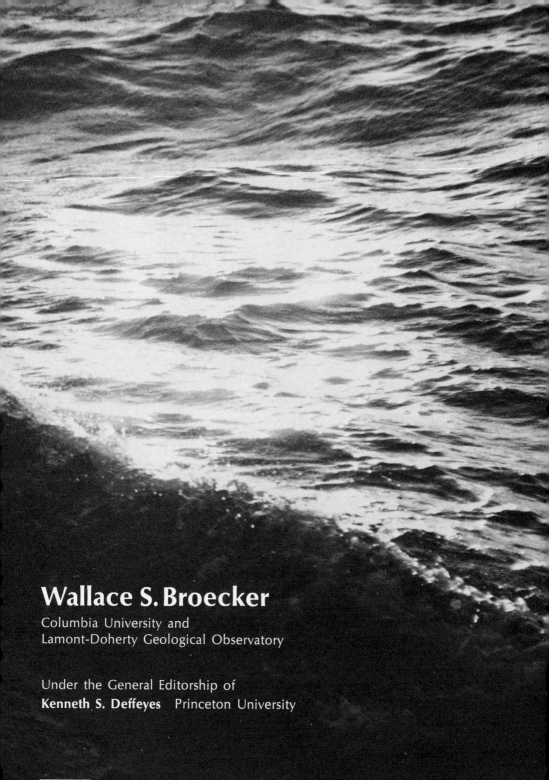

Wallace S. Broecker

Columbia University and
Lamont-Doherty Geological Observatory

Under the General Editorship of
Kenneth S. Deffeyes Princeton University

 Harcourt Brace Jovanovich, Inc.
New York / Chicago / San Francisco / Atlanta

Credits and acknowledgments appear on pages 207–209

CHEMICAL OCEANOGRAPHY

ISBN: 0–15–506437–1

Library of Congress Catalog Card Number: 74–1519

Printed in the United States of America

FOREWORD

Most information comes to us thirdhand or fourth-hand. This is particularly true of scientific information. Original scientific papers are summarized in review articles, then digested into textbooks, and finally outlined by lecturers. Something is lost in each translation, and it is not entirely the fault of students if their examination papers bear little resemblance to the original scientific data.

Scientists seldom have both the inclination and the skill to explain their work directly to students. Wallace S. Broecker is a fortunate exception. In this book, one of the world's best-known oceanographers gives first-year oceanography students clear and simple access to the way he views and understands the sea.

The book began as a series of tapes of Professor Broecker's lectures to his undergraduate oceanography students at Columbia University. The presence of a student audience automatically minimized the possibility of carelessness or inattentiveness on the part of the author. Subsequently, the lecture tapes were transcribed and reworked, and the author then saw the manuscript through three revisions. But the informality of the original lectures survives here.

In the last several years, beginning oceanography textbooks have explained as little as possible about chemical oceanography. Evidently, some authors and editors feel chemical oceanography is more difficult than biological, geological, or physical oceanography. Professor Broecker's book proves that this is not the case.

Wide and detailed in scope yet simple in presentation, *Chemical Oceanography* is useful as a supplement to the beginning oceanography course or as an independent source for self-education. The highest compliment I can pay this book is to confess that I have relied heavily on its various drafts in preparing my own chemical oceanography course.

Kenneth S. Deffeyes

PREFACE

Prior to 1955, chemical measurements in the ocean were made largely as aids to programs designed by physical or biological oceanographers. Consequently, only a very small fraction of the fantastic potential of chemical and isotopic tracers was realized. During the International Geophysical Year (July 1957–December 1958) the atomic technology boom that occurred during the Second World War finally reached the seas. Since then, the field has seen spectacular growth. Great advances have been made in our understanding of the substances dissolved in the sea and buried in its sediments and their utilization as guides to the nature of both past and present processes within the sea. While this book was being written, a very large program, the Geochemical Ocean Section Study (GEOSECS), was initiated, with the objective of mapping on a global scale the chemical and isotopic properties of the sea. These measurements, when completed, will permit the development of a new generation of ocean models that will be far more sophisticated than those given here. Chemical oceanography has at last attained full status as an integral part of the field. Because chemical studies cross so many traditional disciplinary boundaries, the product of the efforts of the new breed of oceanographer will have a profound impact on our understanding of the sea.

In *Chemical Oceanography*, I attempt to summarize the first-order processes taking place within the sea which affect its chemistry. I show that the distribution of chemical species in the water and in the sediment is largely generated by an interaction between mixing cycles and biological cycles. Radioisotope measurements are used to establish the time scale of these processes. The object of this book is to give readers a feeling for the power of the approach rather than to overwhelm them with a mountain of undigested facts.

Despite the great leap forward, for every process within the sea that we think we understand there are dozens for which our knowledge is extremely sparse. Because of this, my attempt to give an overview of the first-order processes at work in the sea is a mixture of fact, speculation, and frank admission of ignorance. Although I have tried to make it clear where I leave the firm ground paved with observations and enter upon the quicksands of intuition, I regret that there is no sure way to make perfectly clear the precise location of this boundary—it is a fuzzy one. In attempts to maximize the use of the measurements made at sea and in the laboratory, one cannot fully resist the temptation to guess what nature is up to. This is true, in essence, of any frontier field.

Many of the ideas presented here were gleaned from papers written by colleagues in the field and from conversations with them. However, the format of the book does not permit more than a casual recognition of their specific contributions. My own research has been greatly stimulated by a close association with a number of graduate students who have written their doctoral theses under my supervision. Such associations are a key ingredient in scientific discovery. Edwin A. Olson, David L. Thurber, Aaron Kaufman, Teh-Lung Ku, Yuan-Hui Li, Michael Bender, Harry James Simpson, Kenneth Wolgemuth, Tsung-Hung Peng, and Steve Emerson all worked with me on problems dealt with in this book. Since my initial involvement in oceanographic research during the International Geophysical Year, I have worked closely with my friend and colleague Taro Takahashi. Most of the expeditions mounted by my group at the Lamont-Doherty Geological Observatory were organized by Ross Horowitz, whose cheerful enthusiasm, attention to detail, and efforts well beyond the call of duty cannot be forgotten. Finally, my wife, Grace, has provided me with a home and an atmosphere conducive to my work. During both my long trips to sea and my short trips into deep thought, she has covered for me in the many details of raising a family.

In the preparation of this book I am indebted to my secretary, Marylou Zickl, who carried the manuscript from the lecture tapes through its many revisions, and to Eleanor Feltser, who converted the hodgepodge of incomplete and redundant sentences on the original typescript into readable English. Tsung-Hung Peng aided with many of the figures, and Harry James Simpson acted as guinea pig by using various drafts of the book in his chemical oceanography course. The enthusiasm of Kenneth Deffeyes pushed me over the initial brink into tape recording my lectures and also sustained me in the harder jobs of editing, illustrating, and proofreading the manuscript throughout its three revisions.

Wallace S. Broecker

CONTENTS

3 HOW FAST DOES THE MILL GRIND? 59

4 THE GREAT MANGANESE NODULE MYSTERY 89

5 ATMOSPHERIC AND VOLCANIC GASES 115

CHEMICAL OCEANOGRAPHY

1

INTERNAL CYCLING

AND THROUGHPUT

The sea is a way station for the products of continental erosion. All substances the sea receives are ultimately passed on to the underlying blanket of sediment. The great tectonic forces that continually alter the geography of the globe eventually bring these sediments above the sea surface and expose them to erosion. Then another trip to the sea begins.

Some of the products of this cycle reach the sea in particulate form. They are dropped by the winds to the sea surface or disgorged by rivers into coastal waters. Within the sea, these rock fragments and soil residues are quite inert. They travel only as far as the currents can carry them before they fall to their burial places on the sea floor.

Of particular interest to us are those products that dissolve during erosion and are carried to the sea in ionic form. They constitute the sea's salt. As long as they remain dissolved, gravity cannot influence them; but other processes at work in the sea ultimately "reprecipitate" the dissolved material delivered from rivers.

The composition of sea salt reflects not only the relative abundance of these dissolved substances in river water but also the difficulty with which they are fixed into sediment forming materials. Sodium, for example, is both abundant in the dissolved matter in rivers and sparingly reactive in the sea. This combination is reflected by its high concentration in sea salt. Calcium, although even more abundant in river water than sodium, is an important ingredient in the shells of marine organisms. Because of its high reactivity, the abundance of calcium in sea water is far lower than that of sodium.

Many components of sea salt show little variation throughout the sea. Others vary in concentration by one-hundredfold from place to place. As we shall see, these differences are largely the result of cycling by organisms. Plants, for example, live only in surface waters, from which they extract certain elements to construct their tissues. Some plant matter is returned to solution after it has been consumed by animals and bacteria, but insoluble and indigestible plant tissues move downward under the influence of gravity—and so the life cycle leads to chemical segregation. On the average, destruction occurs at greater depth than formation. The interaction of this life cycle with the large-scale water circulation pattern in the sea results in the remarkable distribution of these organically influenced elements, not only within the sea itself but also in the sediments forming on the ocean floor.

The aim of this book is to point out the major factors that influence the *average* concentrations of the various components of sea salt and the major factors that produce chemical inhomogeneities in the sea and in the sediments. The approach might be termed "inverse chemical engineering". The ocean is a great chemical plant that processes matter fed to it from rivers and dispenses it as sediment. Unlike most chemical plants, the sea has no advance operational blueprint. As chemical oceanographers, we wander through the plant measuring inputs, losses, internal compositions—trying to reconstruct the missing design. As in most other chemical plants, the two critical features of the ocean are the manner in which its ingredients are stirred and the catalysts that accelerate the rate at which its ingredients react with one another. Oceanic mixing is accomplished by a complex system of currents and a host of turbulent eddies. The catalysts are the enzymes in living organisms. Thus any study of the chemistry of sea water is heavily dependent on knowledge derived from physical oceanographic studies and marine biologic studies.

Beyond a desire to understand the chemistry of the ocean, geo-chemists have a more important goal. Sea salt contains radioisotopes which serve as time clocks. They offer a means of determining the absolute rates of oceanic mixing and of the generation and destruction of plant tissue. However, since the distribution of these time clocks within the sea is controlled by the interaction of physical and biological processes, the influences must be disentangled before the clocks can be read. This difficult task now occupies the attention of many oceanographic researchers.

In this book, we will emphasize the dominant processes operating within the sea. Until these are mastered, it is fruitless to proceed to the less important and much more obscure second-order processes. With this in mind, we explore our first subject—the grand chemical balance existing in the sea.

The salt dissolved in sea water has remarkably constant major constituents. This fact has greatly simplified the task of the physical oceanographer interested in mapping water density patterns within the sea. He needs only to measure the water temperature and one major property of the sea salt (for example, the chloride ion content or the electrical conductivity) to make a very accurate estimate of the density of a given sample of sea water. This task would be extremely complex if the composition of sea salt were more varied.

Yet it is fortunate, too, that compositional constancy is restricted to the *major* components of sea salt and that it does not extend to all the trace components. If this were not the case, the oceanographer would lose one of his most powerful tools, for variations in the trace constituents of sea water are immediate clues to the mixing, biological, and sedimentary processes taking place within the sea.

The major ion matrix of sea salt consists of the following elements: chlorine in the form of Cl^- ion; sulfur in the form of SO_4^{--} ion; and magnesium, potassium, calcium and sodium in the form of Mg^{++}, K^+, Ca^{++}, and Na^+ ions, respectively. These six ions dominate sea salt; their ratios, one to another, are very nearly constant. In fact, only calcium has been shown to vary measurably from place to place in its ratio to the other five elements. Although this constancy also extends to many of the lesser components of sea water (boron, bromine, strontium, fluorine, uranium, cesium, and others), it does not extend to all of them.

Excluding the dissolved gases in sea water, all variations in the composition of sea salt, as far as we know, arise from the removal of elements from surface sea water by organisms and the subsequent destruction of organism-produced particles after downward movement. Deep water masses are richer in the elements consumed by organisms than surface water is. If the ocean were sterile, the chemical composition of sea salt would be almost perfectly uniform. Only slight differences in composition resulting from the transfer of gases between the atmosphere and surface waters of differing temperature would exist.

Plants can live *only* in surface water, where there is enough light to permit photosynthesis. By the time the chemical components of their particulate debris are returned to the water as a result of oxidation or dissolution, downward movement under the influence of gravity or migrating animals has occurred. It is not surprising, then, that the primary chemical differences observed in the ocean are all of the type just mentioned: deep water is enriched relative to surface water.

Figure 1-1 shows the distribution with depth of six important oceanic properties at two locations in the Pacific Ocean. In all cases, the most dramatic change occurs in the upper several hundred meters of the water column. This so-called main thermocline separates the

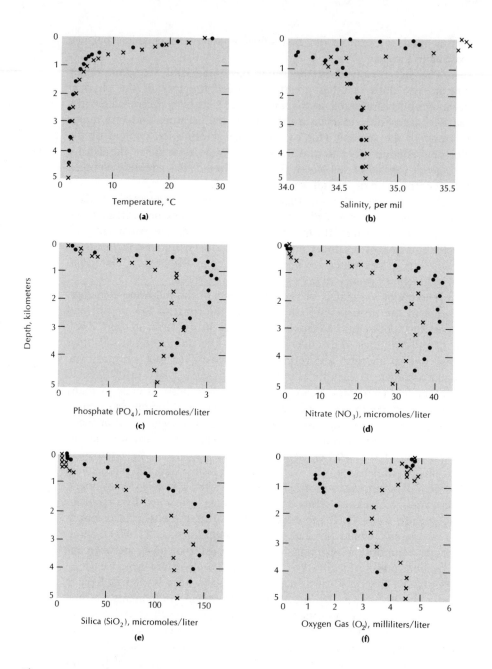

Figure 1-1   The vertical distributions of temperature, salinity, phosphate, nitrate, silica, and oxygen gas for two stations in the Pacific Ocean as determined during the Southern Cross Expedition of the Japanese Research Vessel *Hakuho Maru*. The crosses represent data from a station at 23° S, 170° W; the circles, data from a station at 21° N, 170° W. (Data collected by Yoshio Horibe, Ocean Research Institute, University of Tokyo.)

warm surface waters from the cold underlying waters invading from polar regions. The nutrient elements P, N, and Si show the deep water enrichment mentioned above. Dissolved oxygen gas, by contrast, shows a depletion. Unlike P and N, which are released by respiration, O_2 is consumed. Its pattern is complicated by the fact that cold waters carry with them more dissolved gas than warm waters descending from the surface do. At the southern station, a local excess of evaporation over precipitation gives the surface water a pronounced higher salt content than that found in the underlying cold water. In the northern station, the salinity minimum at about 500 meters depth represents the lateral invasion of Antarctic intermediate water. Keep in mind that these salinity differences are quite small ($\pm.6$ per mil* is equivalent to ±2 percent in salt content) compared to the more than tenfold differences in the concentrations of the three nutrient elements.

Element Classification

About one quarter of the 92 known elements will be considered in this book. They are shown in Table 1-1, grouped as they appear in the periodic table. The first column is comprised of a group of elements known as the alkali metals; they have a valence of +1 (that is, in sea water each atom loses one electron and becomes an ion with a single positive charge). The alkaline earths occupy column (2). These elements have a common valence of +2 (two electrons are given up upon solution in sea water, and the atom becomes a doubly charged ion). The noble gases are in column (8). They do not undergo chemical reactions in sea water, but remain neutral and in gaseous form. Column (7) contains elements with the common property of accepting an extra electron (one of the electrons released by the alkali metal and the alkaline earth metal atoms). These elements are present in sea water with a charge of −1. The elements in columns (3)–(6) all combine with oxygen, and sometimes with hydrogen, to form multiatom ions. In most cases, they form negatively charged ions; in a few cases, they form neutral groupings. Those dissolved units with negative charges are known as *anions*; those with positive charges, as *cations*. Iron, manganese, nickel, cobalt, copper, uranium, and thorium are the only other elements we will consider in this book in addition to those in Table 1-1.

The elements in the table are divided into three major categories: the *biolimiting* elements (those which are almost totally depleted in surface water); the *biointermediate* elements (those which are partially depleted in surface water); and the *biounlimited* elements (those which show no measurable depletion in surface water).

The three known biolimiting elements are nitrogen (N), phos-

---

* A per mil (‰) is a part per thousand; a percent is a part per hundred.

Table 1-1   Abbreviated periodic table showing the elements whose marine chemistry will be dealt with in this book. For each, the dominant ionic and molecular forms found in sea water are noted. For those elements whose distribution within the sea is sufficiently well understood, the designations biolimiting, biointermediate, and biounlimited are given. Asterisks denote minor constituents.

(1)	(2)	(3)	(4)	(5)	(6)	(7)	(8)
Hydrogen H_2O H^{+*} Unlimited							Helium He Unlimited
Lithium Li^+ —	Beryllium Be^{++} —	Boron H_3BO_3 $H_2BO_3^-$ Unlimited	Carbon HCO_3^- CO_3^{--} CO_2 Intermediate	Nitrogen N_2 NO_3^- Limiting	Oxygen H_2O O_2 Intermediate	Fluorine F^- Unlimited	Neon Ne Unlimited
Sodium Na^+ Unlimited	Magnesium Mg^{++} Unlimited	Aluminum † —	Silicon H_4SiO_4 Limiting	Phosphorus $H_2PO_4^{--}$ $H_3PO_4^-$ Limiting	Sulfur SO_4^{--} HS^{-*} H_2S^* Unlimited	Chlorine Cl^- Unlimited	Argon Ar Unlimited
Potassium K^+ Unlimited	Calcium Ca^{++} Intermediate					Bromine Br^- Unlimited	Krypton Kr Unlimited
Rubidium Rb^+ Unlimited	Strontium Sr^{++} Unlimited						Xenon Xe Unlimited
Cesium Cs^+ Unlimited	Barium Ba^{++} Intermediate						Radon Rn Unlimited
	Radium Ra^{++} Intermediate						

† Form in sea water is not known.

phorus (P), and silicon (Si). Plant activity (and, in the case of silicon, animal activity as well) is actually efficient enough to extract these three elements almost totally from surface water. Life in the surface ocean must therefore be limited by the availability of N, P, and Si. Deep water is greatly enriched in these elements in relation to surface water as (c), (d), and (e) in Figure 1-1 indicate. When deep water is returned to the surface, these elements become available to photosynthetic organisms, are fixed into particulate material, and are then carried by gravity back to the deep sea.

Since we have a limited knowledge of the distribution of many of the rare metals dissolved in the sea, elements other than N, P, and Si may eventually be found to be biolimiting, but this is highly unlikely.

To demonstrate that an element is biounlimited, the ratio of that element to the total salt in both surface and deep sea water samples is measured. If the two results are equal within measurement error, the conclusion can be drawn that organisms are not measurably depleting that element from surface water. Those elements currently classified as biounlimited are sodium, potassium, rubidium, cesium, magnesium, strontium, boron, sulfur, fluorine, chlorine, and bromine. After more accurate analyses, one or more of these may be eliminated.

We are certain of only four biointermediate elements: calcium (Ca), carbon (C), barium (Ba), and radium (Ra). The Ca content of surface water salt is about 99 percent that of salt in deep water. The C content of surface water salt is 85 percent that of deep water, and the Ba content of surface water salt is only 25 percent that of deep water.

Of course, most of the oxygen in sea water is in the form of water itself. A small amount occurs in other forms, one of which is oxygen gas (O_2). As this gas is generated by plants and consumed by animals, its anomaly pattern is the reverse of that for other biologically utilized substances—that is, O_2 is most plentiful in surface water and depleted in deep water, as shown in Figure 1-1 (f).

Composition of Particulate Matter

Three major types of particles fall from surface water into deep water: organic tissue, calcium carbonate ($CaCO_3$), and opaline silica (SiO_2). All plants and animals produce organic tissue. Plants extract the ingredients for this tissue from the dissolved salt in sea water, and animals reuse these ingredients by devouring plants and other animals. Many microscopic plants and animals also produce hard parts made of $CaCO_3$ or SiO_2. Hard parts made of $CaCO_3$ are produced by coccolithophorida (plants), foraminifera (animals), and pteropods (animals). Hard parts made of SiO_2 are produced by diatoms (plants) and radiolarians (animals). Photographs of a few of the types of hard parts produced by these organisms appear in Figure 1-2.

The chemical composition of the organic soft tissue formed by plants is relatively constant: for every atom of P in this tissue there are roughly 15 atoms of N and 80 atoms of C. The ratios of these same elements dissolved in deep sea water are 15 atoms of N and 800 atoms of C for every atom of P. If a batch of typical deep Pacific water is brought to the surface, plants will extract the phosphate and the nitrate

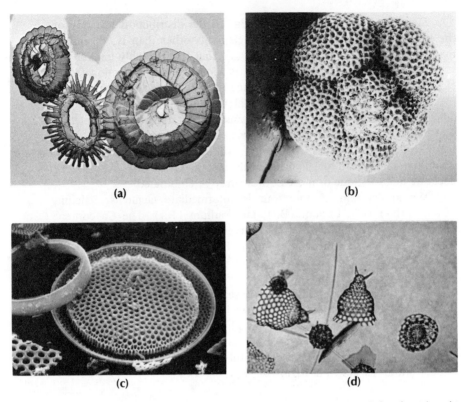

Figure 1-2 Photographs of hard parts formed by (a) coccolithophorida, (b) foraminifera, (c) diatoms, and (d) radiolarians.

they need until they have depleted both elements. Neglecting for the moment the carbon lost to $CaCO_3$, only 80 of the 800 atoms of C will be used. Roughly 90 percent of the carbon will remain in dissolved form. Without phosphate and nitrate, the plants have no use for this remaining dissolved carbon. When animals eat this plant material, they "burn" about 90 percent to obtain energy (releasing the constituent atoms to be redissolved in the water). The remaining 10 percent of the material is used to build animal tissue. Animals require roughly the same P/N/C ratio as plants do. This makes one aspect of ocean chemistry quite simple. To the first approximation, parcels of sea water differ in their P, N, and C contents to the extent that living organisms have removed or added these elements in the fixed organic ratio (see Figure 1-3).

One of the major mysteries of sea water chemistry concerns the P/N ratio. When deep water upwells to the surface, by the time all of its dissolved phosphate has been consumed, by chance or design so has its dissolved nitrate. Why N and P are present in sea water in the same ratio that organisms require them is a little like the chicken and the egg

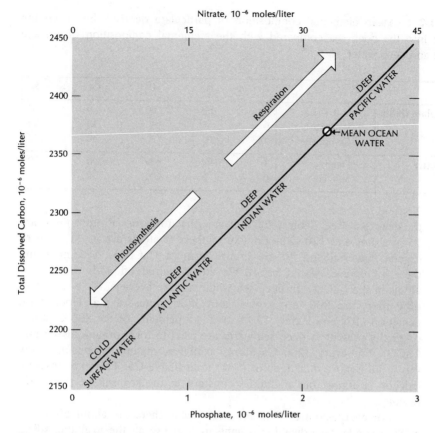

Figure 1-3 Ideal covariance of carbon, nitrate, and phosphate within the ocean. Conditions at the lower (surface water) end of the line are achieved when the limiting nutrients have been exhausted by photosynthesis. The other extreme is fixed by the degree of horizontal enrichment of the nutrient elements within the deep sea. Intermediate values are produced by mixing these end members in varying proportions.

problem. We do not know whether organisms have evolved to use N and P in the 15/1 ratio because that is the ratio in which these elements occur in sea water, or whether organisms have established this 15/1 ratio in the sea through time.

Carbon is also used by organisms to construct their calcium carbonate hard parts. In $CaCO_3$ there is one C atom for every Ca atom. In deep sea water there is one C atom for every four Ca atoms. Thus the generation of $CaCO_3$ depletes carbon four times more rapidly than it does calcium.

On the average, for every two C atoms that fall from the surface sea in the form of organic tissue, one C atom falls in the form of $CaCO_3$. A composite of the organic and $CaCO_3$ debris that fall toward

Table 1-2 Mean elemental composition of particulate debris falling from the surface into the deep sea compared with the elemental composition of average surface and deep water.

		P	:	N	:	C	:	Ca	:	Si
Particulate Debris	Soft Tissue	1	:	15	:	80	:	0	:	0
	Hard Parts	0	:	0	:	40	:	40	:	50
	Composite	1	:	15	:	120	:	40	:	50
Sea Water	Deep	1	:	15	:	800	:	3200	:	50
	Surface	0	:	0	:	680	:	3160	:	0

the deep sea (see Table 1-2) indicates that for every P atom there are 15 N atoms and 120 C atoms (80 of these C atoms are in the form of organic material and 40 are in the form of $CaCO_3$). Accompanying them, of course, would be 40 Ca atoms paired off with the carbon atoms in the $CaCO_3$. In deep sea water, on the average, for every atom of P there are 800 atoms of C and 3200 atoms of Ca. Upon total depletion of N and P only 120 (about 15 percent) of the 800 C atoms originally present in deep sea water are used to form organic tissue and $CaCO_3$. For every 3200 Ca atoms available, organisms consume only 40—a little more than 1 percent of the available Ca. This explains why calcium is almost but not quite constant in the sea salt of the world ocean.

For every atom of P in the deep ocean there are about 50 atoms of Si. Since surface-dwelling organisms also use all the available silica, we can add the appropriate amount of opal to the debris. Because the amount of opal containing 50 Si atoms has nearly the same weight as the amount of $CaCO_3$ containing 40 C atoms and as the amount of organic tissue containing 80 C atoms, the *average* batch of falling particulate matter consists of equal amounts of its three major constituents. At any particular point in the ocean, the proportions might vary significantly from this average because of local ecologic differences in the overlying surface water and because the hard parts ($CaCO_3$ and SiO_2) fall to greater depths before destruction than organic tissue does. (Organic tissue sustains life and is therefore readily consumed; hard parts have no nutrient value and are left to undergo gradual solution.)

Although no direct confirmation has been established, indirect evidence leads us to believe that barium is removed from sea water as barium sulfate ($BaSO_4$). As only 1 atom of Ba need be removed from surface water for every 3000 atoms of C in order to explain the observed depletion in surface water, the presence of $BaSO_4$ in the debris can easily go unnoticed.

The element radium has chemical affinities almost identical to those of barium, and the element strontium (Sr) has affinities quite similar to those of calcium. Thus the formation of $BaSO_4$ and $CaCO_3$ should deplete radium and strontium as well as barium and calcium in surface water. Depletion has been demonstrated for radium but, to date, not for strontium. The expected depletion of Sr is so small as to be beyond the detection limits of presently available techniques.

Factors Controlling the Vertical Segregation of Elements

Now let us be a bit more quantitative about the operation of the ocean's internal cycles. Geochemists who study the sea have found it convenient to treat this immense system as a series of well mixed reservoirs. For our rather elementary look at the ocean, we will divide it into just two such reservoirs—the warm waters and the cold waters (see Figures 1-1(a) and 1-4(a)). We do so because the main obstruction to mixing within the sea is the density difference between the thin skin of warm surface water which covers the tropical and temperate regions of the ocean and the cold water found in the polar regions and throughout the deep sea. The zone separating these two major water types (referred to hereafter as the *oceanic thermocline*) lies between 100 meters (the base of the warm layer) and 1000 meters (the top of the deep water mass) over much of the world ocean. As the poles are approached, the oceanic thermocline rises to the surface and provides a horizontal separation between polar and temperate surface waters. The bulk of the ocean's plant life lives in the upper sunlit portion of the warm water mass. Although the cold reservoir is lighted by the sun, at its polar "outcrops" the area of this exposure is relatively small and the light is absent over much of the year and, when present, is intercepted in part by ice cover. This vast simplification of the ocean allows us to quantify its basic processes rather easily. Once we understand the two-box model shown in Figure 1-4, we will have a first-order picture of how the far more complex *real* ocean operates.

We will make further simplifications. First, we will assume that the only way an element is added to sea water is by river runoff from the continents. We will neglect other means of entry, such as volcanic activity on the floor of the ocean and ground waters running out along the continental margins. Second, we will assume that the only way dissolved salts are removed from the ocean is by the fall of organism-produced particles to the sea floor. Since such particles are not *entirely* destroyed by predators, scavengers, and corrosive waters, a small fraction of this debris is permanently buried in the sediments. Thus rivers add material to the sea, and the particles formed in surface waters fall to the bottom, removing material from the sea. We know that many elements of interest are being replaced in the ocean with great regularity, geologically speaking. For example, the sea is gaining from rivers (and

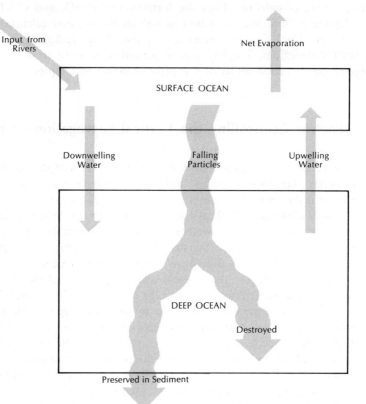

Figure 1-4(a)   Two-box model showing the major fluxes of biologically active elements in the sea. Straight arrows indicate the fluxes of water; wavy arrows, the fluxes of particles.

losing to its sediments) its total content of most of the biologically utilized elements in a period of 10,000–1,000,000 years.

In our model, we will further assume that the operation of the ocean is at *steady state*. By this, we mean that the rates of input and loss of any element to and from the ocean as a whole as well as to and from the warm or the cold reservoirs have remained constant for a sufficiently long time that the concentrations at any point in the sea are not changing with time. The system has stabilized so that gain just balances loss. The content of any element is being steadily renewed, but its amount is not changing. The situation is much like the flow of bottles through a Coca Cola bottling machine. Empty bottles are steadily added at one end; full bottles emerge at the other. The number on the conveyor is always the same. Photographs taken at various times would look alike.

As biologically active elements circulate within the ocean, they

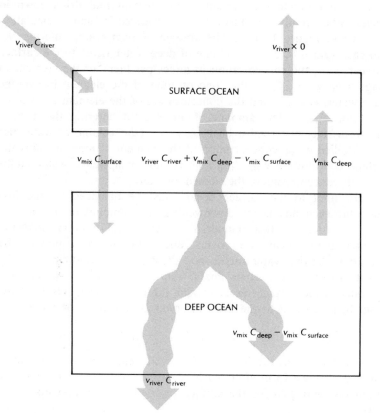

Figure 1-4(b) The same model with the fluxes labeled in accord with the parameters defined in this chapter.

participate in its internal cycles. The warm surface water receives its supply of any given element from two sources: the water entering from rivers, and the water from the deep ocean which is steadily being exchanged with surface water. Every year, a certain amount of deep water upwells to the surface and is incorporated into the warm surface reservoir.

If the concentration of a given element in the surface ocean is to remain constant, then loss by downwelling and particulate settling must exactly match these two inputs. If, as is the case for biologically active elements, the concentration of an element in warm water is less than it is in cold water, then downwelling alone cannot be adequate. For every unit of cold water gained by the warm reservoir, one must be lost to the cold, if the volume of the two oceanic reservoirs is to remain unchanged. Since the water exported to the cold reservoir carries away less of each element than the imported water, particulate loss (plus river input) must account for the difference.

We can estimate the amount of an element that drifts down in particles rather simply. The fluxes to be considered in such a calculation are shown in Figure 1-4(b). The volume of river water entering the ocean each year is v_{river}; the volume of deep water rising to the surface each year, v_{mix}; the concentration in moles per liter* of the element in average deep water, C_{deep}; the concentration of the element in average surface water, $C_{surface}$; and the concentration of the element in average river water, C_{river}. The amount of an element entering the surface ocean from river water must be $v_{river} C_{river}$ (the volume of water per year multiplied by its concentration of the element of interest) plus the upwelling contribution $v_{mix} C_{deep}$ (the volume of upwelled water multiplied by its concentration of the element of interest).

This input to the surface ocean must be balanced by the loss fluxes. One such flux is via downwelling. The amount of an element leaving the surface in this manner is $v_{mix} C_{surface}$. For every elemental unit entering the ocean from rivers, one is lost to the continents by evaporation. As this water carries no salt, it does not participate in the removal process. The particle flux P can be calculated by adding the two inputs to the surface reservoir and subtracting from this total the amount being carried to the deep reservoir by downwelling water:

$$P = v_{river}C_{river} + v_{mix}C_{deep} - v_{mix}C_{surface}$$

If we define g as the fraction of a given element which is removed again in particulate form after reaching surface water, then the balance between loss and gain for the surface reservoir can be reexpressed as follows:

$$g(v_{river}C_{river} + v_{mix}C_{deep}) = v_{river}C_{river} + v_{mix}C_{deep} - v_{mix}C_{surface}$$

Solving for g,

$$g = \frac{v_{river}C_{river} + v_{mix}C_{deep} - v_{mix}C_{surface}}{v_{river}C_{river} + v_{mix}C_{deep}}$$

or

$$g = 1 - \frac{\dfrac{v_{mix}}{v_{river}}\dfrac{C_{surface}}{C_{river}}}{1 + \dfrac{v_{mix}}{v_{river}}\dfrac{C_{deep}}{C_{river}}}$$

As we shall see in Chapter 3, in the present-day ocean the ratio of v_{mix} to v_{river} is about 20; that is, 20 times more water is added to the

---

* There are 6×10^{23} atoms in a mole. One mole of any element has a weight in grams equal to the atomic weight of the element. Thus a mole of C weighs 12 g; a mole of Si, 28 g; and a mole of H, 1 g.

warm reservoir by upwelling than by river runoff. Thus:

$$g = 1 - \frac{20 \dfrac{C_{surface}}{C_{river}}}{1 + 20 \dfrac{C_{deep}}{C_{river}}}$$

For the element phosphorus, C_{deep}/C_{river} is about 5 and $C_{surface}/C_{river}$ is .25 (on the average, river water contains five times less phosphate per liter than deep sea water, and four times more than surface sea water). The corresponding value of g is .95 (95 percent of the phosphate reaching the surface ocean is carried away by falling particles).

On the average, the same amount of any given element must be leaving the sea L as is being added by rivers:

$$L = v_{river}C_{river}$$

Thus, if the removal is accomplished by the burial in the sediments of particles which survive destruction, then the fraction f of a given element carried to the deep sea by the particulate flux P surviving destruction must be equal to the river input:

$$fP = v_{river}C_{river}$$

or

$$f(v_{river}C_{river} + v_{mix}C_{deep} - v_{mix}C_{surface}) = v_{river}C_{river}$$

or

$$f = \frac{1}{1 + \dfrac{v_{mix}}{v_{river}}\left(\dfrac{C_{deep}}{C_{river}} - \dfrac{C_{surface}}{C_{river}}\right)}$$

And since $v_{mix}/v_{river} = 20$:

$$f = \frac{1}{1 + 20\left(\dfrac{C_{deep}}{C_{river}} - \dfrac{C_{surface}}{C_{river}}\right)}$$

For phosphate, f turns out to be .01. (Only 1 percent of the phosphate carried to the deep sea by falling particles survives destruction.)

Now if we multiply the parameter f by the parameter g, we obtain a rather interesting piece of information. Since g tells us the fraction of an element that reaches the surface and is removed in particulate form, and f tells us the fraction of those particles that survive destruction, $f \times g$ gives us the fraction of an element that is removed per oceanic mixing cycle (that is, the transfer from the cold to the warm reservoir).

$$f \times g = \frac{1}{1 + \dfrac{v_{mix}}{v_{river}}\dfrac{C_{deep}}{C_{river}}} = \frac{1}{1 + 20\dfrac{C_{deep}}{C_{river}}}$$

For phosphate, the product of $f \times g$ would be .01 (for f) × .95 (for g), or about .01. Thus 1 percent of the phosphate in the ocean is lost to the sediment during each mixing cycle. If we wait the length of time necessary for upwelling to recycle all the water in the ocean through the surface ocean layer and allow the plants to extract the phosphorus, a loss to the sediment of 1 percent of the phosphate in the ocean will ensue.

As we will see from the C-14 distribution in the ocean (to be discussed in Chapter 3), the time required for one mixing cycle is about 1600 years. Every 1600 years, the entire amount of any given element in the ocean is sent through the surface water mill. To find the time τ required to remove an amount of an element equal to that stored today in the sea, we divide the mixing time T_{mix} by the product of $f \times g$. In equation form:

$$\tau = \frac{T_{mix}}{fg} = \frac{1600}{fg} \text{ years}$$

Since phosphate is removed at the rate of 1 percent every mixing cycle ($f \times g = .01$), the average lifetime of a P atom in the ocean must be about 160,000 years. The calculations for the element phosphorus are summarized in Table 1-3. In words, a typical P atom, upon release from some sedimentary rock by erosion, is carried by rivers to the sea. It then goes through an average of 100 oceanic mixing cycles: 100 times it is fixed by an organism in surface water and becomes part of a particle that sinks and is destroyed in the deep sea. Each time, it waits in the dark abyss about 1600 years before being sent back to the surface. On the average, during the hundredth cycle, the particle bearing the P atom survives destruction and is trapped in the sediment. So our P atom makes 100 round trips of 1600 years each during its stay in the ocean. It then becomes part of the sediment, where it remains for several hundred million years until it is uplifted and exposed again to continental erosion. The life of a typical P atom is indeed bleak. It spends 99.9 percent of its time trapped in the sedimentary rocks of the earth; that is, out of every 200,000,000 years it has only one 160,000 year stint in the ocean. Since the warm layer of surface water is very thin and the time required for particulate loss is small, the P atom spends most of its time in the ocean in the cold dark abyss; for every 1600-year mixing cycle, it spends about four years in the surface water!

The volume of the warm reservoir is only one-twentieth that of the cold. As water resides in the cold reservoir for 1600 years, it must reside in the warm reservoir only one-twentieth as long, or 80 years. Since the concentration of P in the warm reservoir is only 5 percent of that in the cold, the probability that P is removed by particulate matter must exceed the probability that P is removed by downwelling water by a factor of 20. The removal time of P from the warm reservoir must then occur, on the average, 20 times faster than that of its companion water molecules. Thus the residence time of a P atom in the warm reservoir is only one-twentieth of 80 years, or four years.

Table 1-3   Four example calculations for the elements phosphorus, barium, and calcium.

	Phosphorus	Barium	Calcium
1. $f = \dfrac{1}{1 + \dfrac{V_{mix}}{V_{river}}\left(\dfrac{C_{deep}}{C_{river}} - \dfrac{C_{surface}}{C_{river}}\right)}$			
	$\dfrac{1}{1 + 20(5 - .25)} = .01$	$\dfrac{1}{1 + 20(.6 - .2)} = .11$	$\dfrac{1}{1 + 20(25 - 24.75)} = .16$
2. $g = 1 - \dfrac{\dfrac{V_{mix}}{V_{river}}\dfrac{C_{surface}}{C_{river}}}{1 + \dfrac{V_{mix}}{V_{river}}\dfrac{C_{deep}}{C_{river}}}$			
	$1 - \dfrac{20 \times .25}{1 + (20 \times 5)} = .95$	$1 - \dfrac{20 \times .2}{1 + (20 \times .6)} = .70$	$1 - \dfrac{20 \times 24.75}{1 + (20 \times 25)} = .01$
3. $f \times g = \dfrac{1}{1 + \dfrac{V_{mix}}{V_{river}}\dfrac{C_{deep}}{C_{river}}}$			
	$\dfrac{1}{1 + (20 \times 5)} = .01$	$\dfrac{1}{1 + (20 \times .6)} = .07$	$\dfrac{1}{1 + 20 \times 25} = .002$
4. $\tau = \dfrac{T_{mix}}{f \times g}$			
	$\dfrac{2000}{.01} = 2 \times 10^5$	$\dfrac{2000}{.07} = 3 \times 10^4$	$\dfrac{2000}{.002} = 1 \times 10^6$

If we are given the values for f and g for a particular element, we can determine the ratio of its concentration in surface water to that in river water, in deep water to that in river water, and in surface water to that in deep water. To do this, we must solve our equations for the element ratios in terms of f, g, v_{river}, and v_{mix}. The resulting equations are:

$$\frac{C_{surface}}{C_{river}} = \frac{1-g}{fg}\frac{v_{river}}{v_{mix}}$$

$$\frac{C_{deep}}{C_{river}} = \frac{1-fg}{fg}\frac{v_{river}}{v_{mix}}$$

$$\frac{C_{surface}}{C_{deep}} = \frac{1-g}{1-fg}$$

As an example, let's calculate these ratios for phosphorus. Since P has an f value of .01 and a g value of .95,

$$\frac{C_{surface}}{C_{river}} = \frac{1-.95}{.01 \times .95}\frac{1}{20} = 0.25$$

$$\frac{C_{deep}}{C_{river}} = \frac{1-.01 \times .95}{.01 \times .95} \times \frac{1}{20} = 5$$

$$\frac{C_{surface}}{C_{deep}} = \frac{1-.95}{1-.01 \times .95} = .05$$

Having developed this simple chemical model, let us look again at our classification of elements. The biolimiting elements must then be elements for which g is close to unity (they are almost entirely utilized in the surface sea). For the elements in this category (which include P, N, Si), f turns out to be much less than unity (the particles carrying these elements are almost entirely destroyed in the deep sea).

Thus for biolimiting elements:

$$\frac{C_{surface}}{C_{river}} \simeq \frac{1-g}{f}\frac{v_{river}}{v_{mix}} = \frac{1-g}{20f}$$

$$\frac{C_{deep}}{C_{river}} \simeq \frac{1}{f}\frac{v_{river}}{v_{mix}} = \frac{1}{20f}$$

$$\frac{C_{surface}}{C_{deep}} \simeq (1-g)$$

For biounlimited elements, the value of g must be close to zero. For the elements in this category, the following equations must therefore hold:

$$\frac{C_{surface}}{C_{river}} \simeq \frac{1}{fg}\frac{v_{river}}{v_{mix}}$$

$$\frac{C_{deep}}{C_{river}} \simeq \frac{1}{fg}\frac{v_{river}}{v_{mix}}$$

$$\frac{C_{surface}}{C_{deep}} \simeq 1$$

Sodium (Na) is a biounlimited element. Its abundance (relative to the sea salt matrix) is the same in surface and deep water. Sodium is 5000 times more abundant in sea water than in average river water. Thus fg must be about 10^{-5} (only about one part in 100,000 of the Na present in the sea is removed during each mixing cycle). The residence time of sodium in the sea is $100,000 \times 1600$, or 160 million years! In this case, the distinction between f and g has no significance: the ocean gives us only the product of these two parameters.

The first biointermediate element we will consider is barium. On the average, surface ocean water has .18 the Ba concentration of river water and .30 the Ba concentration of deep ocean water. Roughly 70 percent of the barium reaching the surface water is fixed by organisms and sent to the deeps in particulate form. The remaining 30 percent is not needed by the organisms and remains behind for eventual downwelling ($g = .7$). So, although barium shows a depletion in surface water relative to deep water, this depletion is not nearly as great as it is for silicon, nitrate, or phosphate. On the other hand, if we calculate the f value for Ba it is much higher than the f values for these three biolimiting elements. We find that 1 percent of the P remains in the particles and is incorporated into the sediments; for Ba, 11 percent must be retained in the particles incorporated into the sediments. The barium calculations are summarized in Table 1-3.

The ratio of the Ca concentration in surface sea water to that in deep sea water is about .99. The ratio of Ca in the sea to Ca in rivers is 25. Table 1-3 shows the corresponding values of f and g. The results indicate that 16 percent of the $CaCO_3$ formed by organisms is preserved

Table 1-4　Calculation summary for the elements phosphorus, silicon, barium, calcium, sulfur, and sodium.

Category	Element	$\dfrac{C_{surface}}{C_{river}}$	$\dfrac{C_{deep}}{C_{river}}$	g	f	$f \times g$	τ Years
Biolimiting	P	.25	5	.95	.01	$\dfrac{1}{100}$	2×10^5
	Si	.05	1.6	.97	.03	$\dfrac{1}{300}$	6×10^5
Biointermediate	Ba	.20	.60	.70	.11	$\dfrac{1}{13}$	3×10^4
	Ca	30.0	30.3	.01	.16	$\dfrac{1}{500}$	1×10^6
Biounlimited	S	5000	5000	—	—	$\dfrac{1}{10,000}$	2×10^7
	Na	50,000	50,000	—	—	$\dfrac{1}{100,000}$	2×10^8

in the sediments and that the average Ca atom resides in the sea for 500 mixing cycles, or 800,000 years. About 1 percent of the calcium reaching surface water drifts down again in particulate form.

Table 1-4 summarizes the concentration data and corresponding parameters for several of the elements treated in this chapter.

Horizontal Segregation of Elements in the Deep Sea

The simple two-box model of ocean chemistry has shown us the primary features of marine chemistry: the factors controlling the deep with respect to surface concentration and the residence time of a given element in the sea. In the real ocean, the concentration of a given element, while relatively constant from place to place at the surface, shows a considerable geographic variability at great depth. In general, proceeding south in the Atlantic, around Africa, east in the Indian Ocean, and north in the Pacific, the concentration shows a steady increase (see Figure 1-5, showing phosphate distribution). This means that some process taking place within the ocean adds to the vertical segregation of the biologically utilized elements a horizontal component which concentrates them in the deep Pacific Ocean. An interesting characteristic of this horizontal segregation is that the Pacific-ward

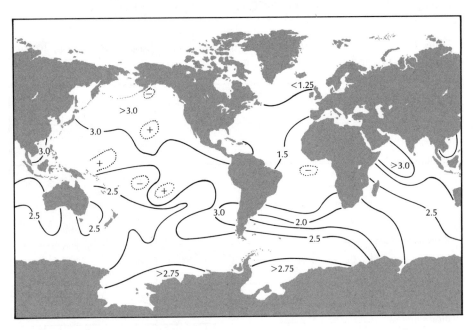

Figure 1-5   Distribution of phosphate at a depth of 2000 meters in the oceans of the world; contour interval, .25 $\times$ 10^{-6} moles/liter.

Table 1-5 Horizontal segregation of biologically utilized elements between the deep Atlantic and the deep Pacific. To permit meaningful calculations for the biointermediate elements carbon and barium, only the excess in deep water over surface water is considered. Note that the horizontal segregation is more pronounced for those elements incorporated into hard parts (silicon and barium) than for those incorporated into soft parts (nitrate and phosphate).

Element	$\dfrac{(C_{\text{deep}} - C_{\text{surface}})_{\text{Pacific}}}{(C_{\text{deep}} - C_{\text{surface}})_{\text{Atlantic}}}$
Nitrogen (as NO_3^-)	2
Phosphorus	2
Carbon	3
Silicon	5
Barium	7

enrichment of the biolimiting elements is different for the hard-part constituents than for the soft-part constituents. Nitrate and phosphate show an enrichment twice as great in the deep Pacific as in the deep Atlantic, whereas silicon shows an enrichment five times as great in the deep Pacific as in the deep Atlantic. Corresponding enrichment factors can be calculated for any biologically utilized element by subtracting its surface water concentration from its deep water concentration and dividing the resultant difference for the Pacific by that for the Atlantic. Values obtained in this way are summarized in Table 1-5.

To understand why these biologically active elements are enriched in the deep Pacific relative to the deep Atlantic, we must understand the deep water circulation pattern in the ocean. At least half of all the water being fed into the deep ocean is generated in the Norwegian Sea at the northern end of the Atlantic Ocean. This water spills over the ridge running from Greenland to the British Isles and flows down the Atlantic, around Africa, through the Indian Ocean, and finally up into the Pacific. There is one complication: joining this North Atlantic Deep Water (NADW) flow toward the deep Pacific is water which has been recooled in the Antarctic Ocean. This cold water sinks from the continental shelf adjacent to the South Atlantic and joins the NADW. It is this mixture that flows through the Indian Ocean into the Pacific Ocean. For the purposes of our discussion, it would be better if the Antarctic recooling did not exist. Thus we will consider a simplified ocean in which the only deep water formation occurs at the north end of the Atlantic.

Oceanographers tell us that the steady production of deep water is balanced by a rather uniform upwelling of water to the surface throughout the world ocean. For every unit of water sinking as NADW one *must* return to the surface. Whereas the sinking is confined to one small portion of the North Atlantic, the return occurs more or less uniformly over the entire ocean.* With this in mind, it is not difficult to see why there is a tendency for a biologically active element to be enriched in Pacific Deep Water (PDW) relative to Atlantic Deep Water (ADW). Plant activity has depleted these elements in surface water sinking into the deep Atlantic. As NADW passes through the Atlantic, particles bearing these biolimiting elements fall into it from the surface and tongues of nutrient-laden water from the deep Indian and deep Pacific oceans recombine with the NADW. Its nutrient content is thus increased. This enrichment process continues as the water moves through the Indian and Pacific deeps. As it proceeds along its path (see Figure 1-6(a)), the current is continually weakened by upwelling. The water lost in this way returns via the surface to the deep water source region (the Norwegian Sea). The nutrients, however, are left behind, and are extracted by organisms and returned to the abyss in particulate form.

The net result is a steady push of the nutrients down and toward the deep Pacific. They can ride the deep current away from the Atlantic but not the returning surface current. This creates an increase in nutrient content along the path of the deep current. This gradient has become so steep that mixing at depth moves the nutrients back toward the Atlantic as fast as they can be exported by the deep current. It is the basic interaction between oceanic circulation and biologic activity that dictates the enrichment of these elements in the deep Pacific relative to the deep Atlantic. A dynamic balance is achieved which stabilizes the Pacific-ward concentration gradient (see Figure 1-6(b)). The relationship between the concentration gradients of the elements C, N, and P was shown previously in Figure 1-3.

An analogy will help you to grasp the essentials of this cycle. Assume that the exhibits in a fun house are located on two levels. The upper floor has a large conveyor belt that moves from right to left; the lower floor, a belt that moves from left to right. Those who enter are free to observe the horrors in any order they wish. There are innumerable escalators from the lower to the upper level. However, there is only one escalator from the upper to the lower level, located at the end of the upper belt. Those who venture to the upper level are harassed by monsters lurking in dark alcoves. These monsters grab the unsuspecting visitors and, after a suitable frightening, drop them through holes to

---

* As we shall see later, it is the average upwelling rate which is nearly the same from ocean basin to ocean basin. Distinct inhomogeneities exist *within* a given basin.

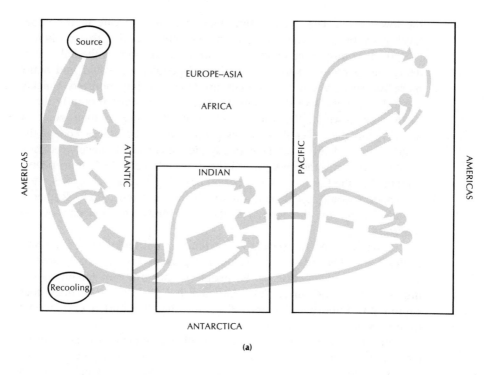

(a)

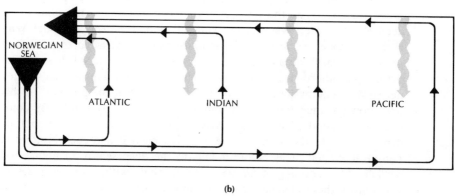

(b)

Figure 1-6 (a) is an idealized map of the patterns of deep water flow (solid lines) and surface water flow (dashed lines). The large circles designate the sinking of NADW in the Norwegian Sea and the recooling of water along the perimeter of the Antarctic Continent; the shaded circles indicate the distributed upwelling which balances this deep water generation. (b) is an idealized vertical section running from the North Atlantic to the North Pacific showing the major advective flow pattern (solid lines) and the rain of particles (wavy lines). The combination of these two cycles leads to the observed distribution of nutrients.

the lower level. The average fun-seeker manages to ride to the upper level many times to view all its mysteries before leaving the fun house.

If, on a busy Sunday afternoon, we were to snap on the lights suddenly and photograph the distribution of people, we would find many more fun-seekers downstairs than up (the untiring monsters quickly track down those who step from the escalator tops and promptly throw them back down to the lower level, relatively close to their point of ascent). On the lower floor, more people would be at one end of the building than at the other (the steady movement of the lower-level belt would carry the wanderers toward the crowded end). As long as people remained in the fun house for a period covering several belt cycles, their distribution would not depend on where they entered or left. Rather it would depend on the speed of the belts and the efficiency of the monsters.

Obviously the fun-seeker is the limiting nutrient and the monster is the plant. The belts and escalators represent the organized flow of water, and the wandering of the people is the turbulent mixing superimposed on this organized flow. Although not perfect, this analogy does capture the important factors influencing the distribution of nutrients in the sea.

As we stated earlier, this horizontal enrichment process is more efficient for some nutrients than for others. Enough phosphate finds its way back along the deep current that the deep Pacific has only twice as much phosphate as the deep Atlantic. But for silica, the content of deep Pacific water averages five times that of the deep Atlantic! Why is this so? One reason is that Si is more rapidly extracted from surface sea water than P, because P is recycled extensively. The soft tissue bearing the phosphorus is in high demand by surface-dwelling animals. They derive their energy by consuming this material, and the products of this oxidation are then returned to solution and reused by plants. By comparing measured plant productivity with estimates of the amount of debris falling to the deep sea, we find that each P and each N atom is reused about ten times before it falls to the deep sea in an undigested particle. The recycling of nitrate and phosphate within the warm reservoir allows these elements to move much further from their point of entry into the surface sea than silica. In other words, N and P can ride the surface current much further back toward the Atlantic than Si can. Another factor contributing to the greater Pacific-ward enrichment of silica is that opal falls deeper in the sea before being destroyed than organic tissue does. The deeper injection of opal more effectively isolates silica from the return flow of near-surface waters. This latter effect is reflected in the vertical distribution of the elements in the Pacific Ocean (see Figure 1-1). Tissue-derived elements show a mid-depth maximum. Elements derived from hard parts increase all the way to the bottom.

Summary

In this chapter, we have learned that the known chemical differences within the ocean can be explained on the basis of the fixation of elements into particulate phases by organisms dwelling in surface water and the subsequent release of these elements back to solution after the particles have been carried by gravity or animals to deeper zones in the ocean. This produces a surface-to-deep enrichment in the biologically active elements. In addition to this vertical segregation, there is also a horizontal segregation within the deep ocean: elemental concentrations are greater in the deep Pacific than in the deep Atlantic. This Atlantic-to-Pacific enrichment is related to the dynamics of deep water circulation. Deep water generated at the north end of the North Atlantic flows southward and then around Africa into the Indian and Pacific oceans. Upwelling occurs more or less uniformly over the entire ocean. The superposition of the particle cycle on this circulation pattern pushes the biologically active elements toward the deep Pacific. Superimposed on this internal cycle is a gradual loss of biologically active elements from the sea. The residence time of such elements in the sea depends on the degree to which they are incorporated by organisms during each pass through the warm surface reservoir and the degree to which the particulate matter so formed survives destruction in the deep sea.

This model emphasizes only the major processes at work in the sea. The role of inorganic particles as removal agents has not been discussed. Second-order features in the global circulation pattern have not been mentioned. Deviations from the average ratio of one nutrient to another during uptake and release have not been stressed. A complete picture of the ocean must, of course, include these factors. However, as these lesser processes are very complex and, at present, poorly understood, they are not subject to the broad treatment given here. Nevertheless, they cannot be ignored. They may hold surprises even to the extent that opinions presented here as to what the dominant processes are might require revision.

SUPPLEMENTARY READINGS

■ Standard textbooks covering the broad range of subjects included in the field of oceanography:

Sverdrup, H.U., Johnson, M.W., and Fleming, R.H. *The Oceans, Their Physics, Chemistry, and General Biology.* Englewood Cliffs, N.J.: Prentice-Hall, 1942. The classic oceanography textbook. Although written before World War II, it still gives the best treatment of the foundations of the discipline.

Weyl, P.K. *Oceanography: An Introduction to the Marine Environment.* New York: John Wiley & Sons, Inc., 1970. A recently written and quite readable book briefly treating a broad range of subjects.

Turekian, K.K. *Oceans.* Foundations of Earth Science Series. Englewood Cliffs, N.J.: Prentice-Hall, 1968. A compact treatment of oceanography from the viewpoint of a geochemist.

■ Other textbooks in marine chemistry:

Riley, J.P., and Skirrow, G., (eds.). *Chemical Oceanography,* Volumes 1 and 2. London and New York: Academic Press, 1965. A multiauthored book covering a broad range of subjects in chemical oceanography. Most contributions review the scientific literature without a concerted attempt to explain how the system operates. A key to the professional literature.

Riley, J.P., and Chester, R. *Introduction to Marine Chemistry.* New York: Academic Press, 1971. This text nicely complements the material in Chapter 1 by presenting far more of the actual data collected by workers in the field.

Horne, R.A. *Marine Chemistry.* New York: John Wiley & Sons, Inc., 1969. A treatment of chemical oceanography emphasizing the thermodynamic and physical-chemical aspects of the subject.

■ Books devoted to the chemistry of continental runoff:

Garrels, R.M., and Mackenzie, F.T. *Evolution of Sedimentary Rocks.* New York: W. W. Norton & Company, Inc., 1971. An excellent elementary treatment of the processes at work in denuding the continents and of the geologic record this activity generates in the sediments.

Livingston, D.A. *Chemical Composition of Rivers and Lakes.* U.S. Geological Survey Paper 440G, 1963. A compilation of chemical measurements made on continental waters.

Stumm, W., and Morgan, J.J. *Aquatic Chemistry.* New York: John Wiley & Sons, Inc., 1970. A systematic treatment of the physical chemistry of processes occurring in fresh waters.

PROBLEMS

1-1   Element x has an average concentration of 2 milligrams per liter (mg/l) in river water, 4 mg/l in surface water, and 16 mg/ in deep sea water. Is this element biolimiting, biointermediate, or biounlimited? What are the values of g, f, and τ for this element? Where would you find water richest in this element?

1-2   Element y has an f value of .20 and a g value of .50. Its concentration in average river water is 10 micrograms per liter (μg/l). What is its average concentration in the deep sea? In the surface sea? What is its average residence time in the sea?

1-3 Marine organisms use Sr as well as Ca to build their $CaCO_3$ shells. The ratio of Sr to Ca in their shells is only one-fifth that dissolved in the water in which they dwell. If this is the only means by which Sr is removed from sea water, what should be the ratio of Sr in surface water to Sr in deep water of the same salinity?

1-4 If 1 mole of O_2 is required to "burn" an amount of organic tissue containing 1 mole of C, then how many moles of O_2 would disappear from a sample of deep water for each mole of P added by respiration?

1-5 If, at some time in the past, deep water had formed at the north end of the Pacific rather than at the north end of the Atlantic, how would the chemistry of the sea have been different from what it is today?

1-6 If organisms were to become only half as efficient as they are in consuming the organic debris falling to the deep sea (and this were to remain true for several hundred thousands of years), how would the chemistry of the sea change?

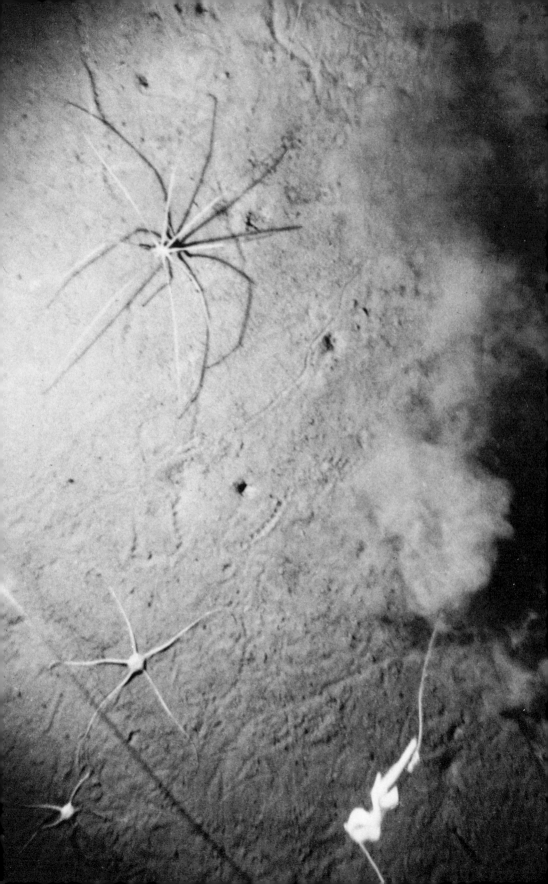

2

THE SEDIMENTARY

IMPRINT

The character of the sediments collecting at any point on the sea floor is closely related to the distribution of the biologically utilized elements within the sea. Much of our knowledge of the chemical processes taking place within the sea stems from studies of these sediments. They are, in the final accounting, the direct or indirect evidence of the inputs to the sea and of the recycling within the sea.

Sediment Types

Deep sea sediments consist of three major components. *Detrital* material (see Table 2-1) is derived mainly from the erosion of the world's continents. Most of this material is in the form of alumino-silicate minerals dropped by winds onto the oceans or carried by river waters past their mouths. It is spread over the entire ocean floor, in greatest abundance along continental margins and in lesser amounts on the deep floor remote from land. Deep sea cores taken from anywhere except within a few hundred kilometers of land masses show sedimentation rates on the order of .3 grams of alumino-silicate debris per square centimeter (g/cm^2) of ocean floor per 1000 years of accumulation time. A range

Table 2-1   Mineralogy of alumino-silicate detritus swept into the sea from the continents via rivers and air.

Mineral	Chemical Composition
Quartz	SiO_2
Orthoclase	$KAlSi_3O_8$
Plagioclase	$xNaAlSi_3O_8 + (1-x)CaAl_2Si_2O_8$
Kaolinite	$Al_2Si_2O_5(OH)_4$
Illite	$KAl_3Si_3O_{10}(OH)_2$
Montmorillonite	$Al_2Si_4O_{10}(OH)_2 \cdot xH_2O$
Chlorite	$Mg_5Al_2Si_3O_{10}(OH)_8$

of about a factor of 3 on either side of this mean rate is broad enough to include a large fraction of the sea floor.

The *authigenic* minerals in sediments are of a more immediate origin. They are formed in place by spontaneous crystallization either on the sea floor or within the sediment column, and they make up a very small fraction of the total sediment. (We will postpone our consideration of these minerals until Chapter 4.)

Organic components are represented by two important substances that are produced as hard parts of organisms. These materials are calcite and opal, precipitated by free floating plankton in the waters of the surface ocean. As mentioned in Chapter 1, calcite is formed by coccoliths (plants) and forams (animals); the opal, by diatoms (plants) and radiolarians (animals). The distribution of calcite and of opal in the ocean sediments is not at all uniform. In some places in the ocean we find an enormous amount of opal mixed with the ever present detrital alumino-silicates (red clay). Such sediments can run as high as 90 percent opal. In other places, we find that the clay blanket is diluted with an enormous amount of $CaCO_3$. Sediments as high as 90 percent $CaCO_3$ are not uncommon. In a few places, calcite and opal occur together. In many others, we find neither of these substances; the alumino-silicate debris accumulates alone.

Organic tissue itself rarely makes an important contribution to deep sea sediments (a tribute to the efficiency of the animals and bacteria that feed on it). Its distribution is fairly uniform (on the order of .3 percent of the total sediment). But since the number of analyses that have been made of the tissue debris in the sediment is not large, we do not have a good understanding of the details of the accumulation pattern for organic debris. All we can say is that the little bit of organic material that is not destroyed in deep water is fairly well mixed with the bulk of the sediment (unlike the opal and calcite, which are sometimes absent and other times dominant).

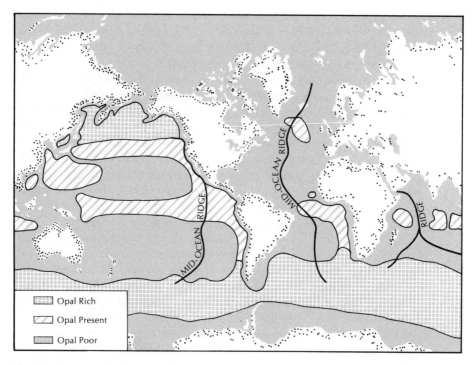

Figure 2-1 Map showing the distribution of SiO$_2$-rich sediments on the sea floor. (Unpublished data from K. Venkatarathnam and James Hays, Lamont-Doherty Geological Observatory.)

Distribution of Opal (SiO$_2$)

The distribution of opal in deep sea sediments reflects those areas where diatoms and radiolarians live in great abundance in the overlying waters (see Figure 2-1). The entire ocean, including the deep ocean, is well undersaturated with respect to the mineral opal (that is, if left long enough in sea water, opal should spontaneously dissolve). In theory, then, opal should not be found in the sediments. Yet roughly 3 percent of the opal grains become a permanent part of the sediment, indicating that the remains of certain species of diatoms and radiolarians are more inert than others. These organisms may form an opaline test that is neatly armored with resistant organic material, or they may have a much lower surface area/volume ratio. In any case, certain species (especially the thick-walled ones) are preferentially incorporated into the sediment. The material found in net tows made in the surface ocean contain a number of species of diatoms and radiolarians that is an order of magnitude greater than the number found in the sediments underlying these same areas. The very fragile species dissolve totally during their trip down through the water column or soon after they reach the bottom.

More rugged species are able to withstand the corrosive bottom water and are preserved. So far as we can tell, the opaline contribution in the sediment reflects mainly the overlying productivity of opaline material. Lateral redistribution by currents is unimportant.

If the upwelling of deep water is uniform over the world ocean, the rate of production of opal should increase gradually around the ocean in an Atlantic to Pacific direction in accord with the gradually increasing silica content of upwelling water. Although, on the average, the sediments do become more siliceous from Atlantic to Pacific, the distribution of silica in sediments from a given ocean basin is far from uniform. Sediments with high opal accumulation rates underlie areas where there is *enhanced* upwelling of deep water. Such zones exist along the equator, at the boundaries between the temperate and polar seas, and along certain continental margins. When silica reaches the surface in these zones, it is not permitted to migrate very far beyond the limits of the upwelling area because diatoms and radiolarians quickly fix it into particles and send it back down to the depths. Opaline-rich sediment thus formed is especially prominent in the equatorial Pacific, the Antarctic, and the northern North Pacific. Enhanced upwelling occurs in the Atlantic, too, but the silica-starved waters of this ocean are not capable of supporting enough opal production to make it a major component in the sediment.

Quantitative opal contents in deep sea sediments are not easily measured. Most of the data concerning the distribution of this substance are compiled from the qualitative evidence supplied by paleontologists. In fact, one of the major remaining studies of marine sediments is the accurate analysis of the opal content of existing sediment collections.

Distribution of Calcite (CaCO₃)

The distribution of $CaCO_3$ in sediments differs from that of opal (see Figure 2-2). First, the pattern of production of $CaCO_3$ hard parts by organisms in surface waters is more uniform. The abundance of the organisms producing $CaCO_3$ is limited by the availability of nitrate and phosphate for their soft tissue (calcium and carbon are never depleted). The recycling of phosphate and nitrate *within* the surface sea permits these two elements (which are essential to the $CaCO_3$-producing organisms) to travel much further from their point of upwelling than Si does. More important to the distribution of calcite than productivity differences in the overlying surface waters, however, are the chemical differences in the deep waters in contact with the sediment. Unlike the situation for opal, the ocean is only locally undersaturated with respect to (or corrosive toward) $CaCO_3$. Those $CaCO_3$ hard parts that fall into deep water that is undersaturated with respect to $CaCO_3$ are largely

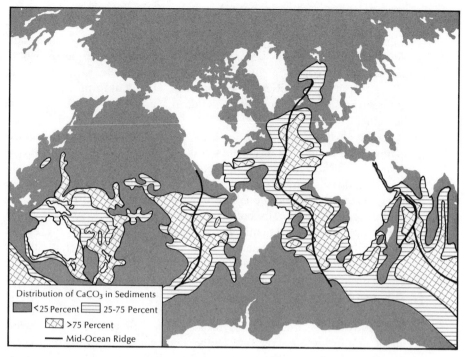

Figure 2-2   Map showing the distribution of CaCO₃-rich sediments on the sea floor. (The Atlantic and Indian distributions are based on data obtained by Pierre Biscaye, K. Venkatarathnam, James Gardner, and Thomas Kellogg, all of Lamont-Doherty Geological Observatory, Karl Turekian, Yale University, and David Ellis, Oregon State University. The Pacific distribution is based on data compiled by A. P. Lisitzin from the USSR.)

dissolved, and the $CaCO_3$ remains that fall into supersaturated deep water are preserved. Recent laboratory studies at Yale University show that the rate of attack by undersaturated waters does not become significant until less than half the saturation carbonate ion content is achieved. So perhaps we should say that those $CaCO_3$ hard parts falling into waters whose degree of undersaturation lies beyond this 50-percent level are largely destroyed; those falling into water undersaturated to less than this extent are only partially destroyed; and those falling into supersaturated waters are preserved. Indeed, consistent with this refinement, depth zones of partial dissolution are often found in the sediment. Later we will relate these observations to the Ca cycle described in Chapter 1. (We found that only 16 percent of the $CaCO_3$ formed during each mixing cycle was preserved in the sediments; the remainder dissolved.)

Before explaining why this is true, we must point out that not just the mineral calcite (the stable form of $CaCO_3$ at any temperature and pressure found within the ocean) is precipitated by marine organisms, but that aragonite (a metastable form of $CaCO_3$ under oceanic conditions) is also formed by one group of marine animals—pteropods. Since all ocean surface water is supersaturated with respect to both calcite and aragonite, a given organism can generate whichever form of $CaCO_3$ it finds easier to precipitate. The two forms show quite different preservation patterns in the deep sea which provide an important clue.

The shells of pteropods in the Atlantic Ocean are found in sediments down to about 2500 meters in depth, but not in sediments deeper than this. In the Pacific Ocean, pteropod shells are found only in sediments shallower than a few hundred meters. Calcite forms disappear (or show evidence of extensive attack by corrosive water) at much greater depths than aragonite in both oceans. Substantial calcite dissolution effects appear in the Atlantic at about 5000 meters and in the Pacific at a depth of about 3500 meters. This distribution suggests that Atlantic waters below 5000 meters are sufficiently undersaturated with respect to calcite and Atlantic waters below 2500 meters are sufficiently undersaturated with respect to aragonite to permit almost total dissolution of these substances. In the Pacific, waters below a few hundred meters are sufficiently undersaturated with respect to aragonite and waters below about 3500 meters are sufficiently undersaturated with respect to calcite to destroy the major influx of these materials. As we will see, the shallower preservation depth in the Pacific relative to the Atlantic is a natural consequence of the horizontal gradient in chemical properties outlined in Chapter 1.

Degree of CaCO₃ Saturation

To understand the relationship of the distribution of $CaCO_3$ in sediments to the distribution of biologically utilized elements within the sea, we need to examine the designations *saturation, undersaturation,* and *supersaturation* more meaningfully. For any salt (or mineral) placed in contact with sea water, the product of the concentrations of the ions formed during its solution will reach a limit at which the solution has become saturated with the given salt (or mineral). This limit depends on the temperature and pressure (the depth) of the sea water. In the case of $CaCO_3$, calcium ions and carbonate ions are formed. The degree of saturation D is written as:

$$D = \frac{([Ca^{++}] [CO_3^{--}])_{\text{sea water}}}{([Ca^{++}] [CO_3^{--}])_{\text{sea water saturated with calcite}}}$$

where the brackets indicate the concentration of the enclosed ion. If calcium and carbonate ions are present in sea water beyond the saturation product, then the water is supersaturated with respect to calcite. If this solution is left undisturbed long enough, $CaCO_3$ will crystallize spontaneously until the saturation product is established (until D becomes unity). However, the spontaneous precipitation of $CaCO_3$ takes place very slowly in sea water: apparently, the presence of magnesium ions in the water strongly impedes the growth rate of $CaCO_3$ crystals. Thus the only important way $CaCO_3$ is removed from the sea water is by the action of organisms. In their absence, sea water would be highly supersaturated simply because the spontaneous processes that seek to reestablish saturation occur so slowly. A *high* degree of supersaturation would be required to rid the ocean of $CaCO_3$ as fast as it was supplied by rivers.

Obviously, working with a collected ocean water sample alone will give us only the values in the numerator of the expression for D shown on page 36. It will not tell us the degree of saturation of $CaCO_3$ for the sample. To find the denominator, we must add crystals of calcite (or aragonite) to a sterile sea water solution maintained in the laboratory and leave the sample long enough for the water and the solid to achieve equilibrium. This is the only way we can determine the saturation product. Once it is established (at various combinations of temperature and pressure), then the ratio of the oceanic ionic product to the ionic product for laboratory equilibrium gives D. If $D = 1$, the sea water is exactly saturated; if $D > 1$, the water is supersaturated; and if $D < 1$, the water is undersaturated.

For $CaCO_3$, the equation can be simplified. The calcium content of sea water remains nearly constant, varying with respect to the salt matrix only over a range of 1 percent and as a result of salinity differences by only another ± 1 percent.* Since the variations in CO_3^{--} ion are far larger by comparison, we can assume that the Ca^{++} ion content is the same everywhere in the sea and in the sea water samples used in the laboratory, canceling it out in the numerator and denominator of the expression for D and yielding:

$$D = \frac{[CO_3^{--}]_{sea\ water}}{[CO_3^{--}]_{saturated\ sea\ water}}$$

D is then the simple ratio of the carbonate ion content in the sea water sample to the carbonate ion content in the sea water solution saturated in the laboratory.

But the saturation point varies with temperature and pressure, depending on whether we are dealing with the mineral calcite or with the mineral aragonite. Because of its metastability, aragonite is more soluble than calcite; thus the equilibrium carbonate ion content for

---

* The evaporation–precipitation cycle results in salinity differences from place to place in the sea (see Figure 1-1(b)). Of course, the salt is left behind when water evaporates, enriching the residual sea water.

Table 2-2   Saturation carbonate ion content of sea water as a function of temperature and pressure for the minerals calcite (precipitated by coccoliths and forams) and aragonite (a metastable polymorph of $CaCO_3$ precipitated by pteropods).

Temperature, ° C	Pressure, atm*	Saturation Carbonate Ion Content, 10^{-6} moles/liter†	
		Calcite	Aragonite
24	1	53	90
2	1	72	110
2	250	97	144
2	500	130	190

* A pressure of 100 atm is achieved at close to 1000 meters depth in the sea.
† 10^{-6} moles/liter (one micromole per liter) is equivalent to 10^{-3} moles/m^3 (1 millimole per cubic meter).

aragonite will always be higher under corresponding conditions of temperature and pressure than that for calcite (see Table 2-2).

The temperature range for most of the ocean extends from about 1° C in deep and polar waters to about 30° C in tropical surface waters; the pressure range extends from 1 atmosphere (atm) at the surface to 550 atm at a depth of 5500 meters (the maximum bottom depth for much of the sea). Now $CaCO_3$ is an unusual salt: the colder the water, the more soluble it becomes! The difference in the saturation concentration of CO_3^{--} ions in a surface water sample from the tropics, where the temperature averages 24° C, and a surface water sample from the polar regions, where the average temperature is 2° C, is about 25 percent (see Table 2-2). The solubility of $CaCO_3$ also increases with pressure, and pressure increases with depth. All the water in the deep ocean is cold. Hence there is no point in considering what happens at a high pressure and a high temperature of 24° C; we only need to know what happens at 2° C. Table 2-2 shows that both calcite and aragonite are almost twice as soluble at 500 atm (2° C) as they are at 1 atm (2° C). The pressure dependence clearly dominates the temperature dependence. Moreover, since the temperature below 1000 meters is nearly constant, the variation in solubility of $CaCO_3$ within the deep water mass is entirely the result of the rise of pressure with depth.

Variations in the Carbonate Ion Content of Sea Water

Only one question remains. How does the carbonate ion concentration vary from place to place within the sea? With this information in hand, we can see how the degree of supersaturation of sea water varies with

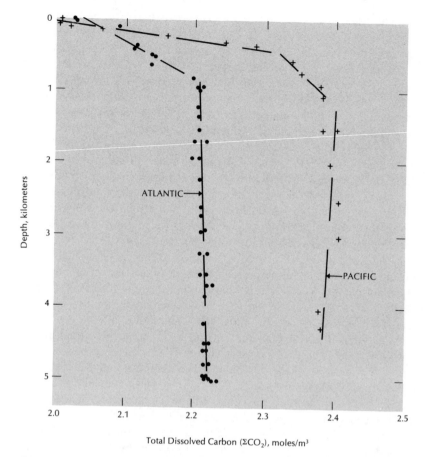

Total Dissolved Carbon (ΣCO_2), moles/m³

Figure 2-3 Variation of total dissolved inorganic carbon content (ΣCO_2) with depth in the Atlantic (36° N, 68° W) and in the Pacific (28° N, 122° W). (Data collected by Raymond Weiss, Scripps Institution of Oceanography.)

respect to the $CaCO_3$ minerals calcite and aragonite. To determine this, we must probe more deeply into some of the chemical reactions taking place in the ocean.

Obviously, the CO_3^{--} ion content of sea water varies with the total amount of dissolved carbon. Figure 2-3 shows the systematic vertical variation in the total dissolved carbon content (ΣCO_2) of sea water. We already learned in Chapter 1 that the lowest amount is found at the surface; this value represents the carbon left after all the limiting nutrients have been used. Atlantic Deep Water shows a carbon excess of about 10 percent over warm surface water; Pacific Deep Water shows an excess of about 20 percent over warm surface water. As we will see, the excesses over cold surface water in both oceans are somewhat smaller.

Variations in the Carbonate Ion Content of Sea Water / **39**

The fraction of the total dissolved carbon that is in carbonate ion form also varies from place to place in the sea. Dissolved carbon is present in the ocean in three forms: dissolved carbon dioxide gas (CO_2), bicarbonate (HCO_3^-) ion, and carbonate (CO_3^{--}) ion. Carbon is one of the few elements in the ocean that can exist in ions having different charges: as HCO_3^- ion, it has a charge of -1; as CO_3^{--} ion, it has a charge of -2. (The amount of gaseous CO_2 in the ocean is very small and will therefore be eliminated from this discussion.)

The ratio of CO_3^{--} to HCO_3^- varies from place to place in the sea and bears a definite relation to the sea's overall balance of charge. The major contributors of positively charged ions (cations) in the sea are Na, K, Mg, and Ca (see Table 2-3). To find the total number of units of positive charge per unit volume of sea water (for convenience, we use the cubic meter m^3 or 1000 liters), we multiply the moles/m^3 of each cation by its charge and total the results: 470 units of charge from Na, 10 from K, 106 from Mg, and 20 from Ca. This adds up to 606 units of positive charge per cubic meter. Since sea water can have no overall charge, its negatively charged ions (anions) must add up to exactly the same value. If we take the three major anions—chloride (Cl), sulfate (SO_4), and bromide (Br)—in moles/m^3, we find that their charges add up to 604 units of negative charge per cubic meter, so we are short 2 units of negative charge. This difference between the sum of the negative charges is balanced in the ocean by the dissolved carbon ions HCO_3^- and CO_3^{--}. The ratio of these two anions to one another varies in such a way that charge is balanced. When more negative charge is needed to balance the cations present, HCO_3^- is converted to CO_3^{--}; when less is needed, CO_3^{--} is converted to HCO_3^-. The amount of excess positive charge balanced by bicarbonate and carbonate ions is a small but, as we shall see, very important residual that holds the key to the carbonate ion distribution within the sea.*

In our discussions we will neglect the element boron. The presence of boron in sea water requires a small correction in all the calculations outlined here, for it also exists in two ionic forms with different charges ($H_2BO_3^-$ and H_3BO_3). Its inclusion here would make everything far more complicated without greatly altering the result. To avoid this complexity, we will treat the ocean as if it did not contain boron.

To understand what determines the ratio of carbonate ions to bicarbonate ions in a given unit of sea water, we must recognize two restrictions. First, the total amount of dissolved carbon present in the water cannot be changed as a result of our manipulation of the charges. There are a fixed number of carbon atoms and they must exist in either

---

* Chemists measure alkalinity in units of equivalents/liter. An equivalent is equal numerically to a mole. A mole of Na^+ ions would carry an equivalent of positive charge; a mole of SO_4^{--} ions would carry two equivalents of negative charge. In this book we will simplify matters by using moles of charge rather than equivalents as the unit of alkalinity.

Table 2-3 Charge balance in sea water: the excess cation charge is balanced by the dissociation of carbonic acid (H_2CO_3) into bicarbonate (HCO_3^-) and carbonate (CO_3^{--}) ions.

Positive			Negative		
Cation	Mass, moles/m³	Charge, moles/m³	Anion	Mass, moles/m³	Charge, moles/m³
Na+	470	470	Cl−	547	547
K+	10	10	SO₄−−	28	56
Mg++	53	106	Br−	1	1
Ca++	10	20			
Σ	—	606	Σ	—	604
			HCO₃− + CO₃−−	—	2
			Σ′	—	606

bicarbonate or carbonate form. Second, the sum of all the bicarbonate and carbonate charges must just balance the excess cations so that the sea water maintains its electrical neutrality. Respectively, these two restrictions can be stated mathematically as:

$$[\Sigma CO_2] = [HCO_3^-] + [CO_3^{--}]$$

where ΣCO_2 is total dissolved carbon, and

$$[A] = [HCO_3^-] + 2[CO_3^{--}]$$

where A is the alkalinity of the water (that is, the excess positive charge to be balanced by CO_3^{--} and HCO_3^- ions). The alkalinity is given by:

$$[A] = [Na^+] + [K^+] + 2[Mg^{++}] + 2[Ca^{++}] + \cdots$$
$$- [Cl^-] - 2[SO_4^{--}] - [Br^-] - \cdots$$

The concentrations of doubly charged ions are multiplied by two in order to give the concentrations of the charges they carry.

If we subtract the ΣCO_2 equation from the first A equation, we find that:

$$[CO_3^{--}] = [A] - [\Sigma CO_2]$$

Thus the carbonate ion content of any unit of sea water sample is equal to its alkalinity (excess positive charge) minus its total dissolved carbon content. Substituting this result in either of the original equations, we obtain:

$$[HCO_3^-] = 2[\Sigma CO_2] - [A]$$

Here we see that the bicarbonate ion content equals twice the total dissolved carbon content minus the alkalinity.

Table 2-4 Impact of particulate production on those properties of sea water related to the solubility of $CaCO_3$ (total dissolved carbon content ΣCO_2, calcium content Ca^{++}, and alkalinity A).

Decrease in Dissolved Property, moles/m³	Detrital Component Formed			
	Organic Tissue, 2 moles	Calcite, 1 mole	Opal, 1 mole	Aggregate, 4 moles
$\Delta[\Sigma CO_2]$	2	1	0	3
$\Delta[Ca^{++}]$	0	1	0	1
$\Delta[A]$	0	2	0	2

So to understand how the carbonate ion content of sea water varies from place to place, all we need to know is how the alkalinity of the water and how the total dissolved carbon content of the water vary.

As we have already seen, the total dissolved carbon content of sea water varies because plants extract material from water and then their remains (or the remains of the animals they support) sink into the deep water and are largely destroyed. *Two* processes are at work here: organic tissue formation and $CaCO_3$ or hard-part formation. The formation of *organic* tissue utilizes carbon and hence changes the ΣCO_2 of the water. It has, however, no effect on the alkalinity of the water because none of the ionic concentrations appearing in the equation defining A change.

Removal of $CaCO_3$ from the water changes both the total dissolved carbon and the alkalinity. Organisms use carbon to form $CaCO_3$ and thus remove it from the water. The alkalinity of the water changes because when $CaCO_3$ is formed in the water, Ca ion—one of the contributors to the net positive charge—is removed. Each mole of $CaCO_3$ formed results in the extraction of 2 moles of positive charge (a mole of Ca^{++}); thus the creation of $CaCO_3$ changes the alkalinity of sea water by just twice the amount that it changes the carbon content.*

In Chapter 1, we estimated that for every two C atoms that are removed from surface water as organic tissue one C atom is removed as $CaCO_3$. Table 2-4 shows the total effect on sea water of adding or subtracting organic debris plus $CaCO_3$. From the table, we see that for every three C atoms used (two in organic material and one in $CaCO_3$) one Ca atom is used. This uses up two positive charges and correspondingly reduces the alkalinity. For the combined organic–$CaCO_3$ debris, the change in alkalinity is equal to two-thirds the change in total dissolved carbon. If the composition of the debris being formed and

---

* The third type of particulate material being formed—opal—affects neither the alkalinity nor the total carbon content of the water.

destroyed were constant everywhere, then the carbonate ion content of sea water would follow some very simple rules. Although this is not quite the case, a good first approximation of how the system operates can be made by assuming that it is.

To see how the CO_3^{--} ion content varies, let us start with surface water that is phosphorus-free and cold ($\sim 2° C$). Remember that this water contains the *minimum* amount of carbon, since as much carbon as possible is removed when available phosphorus is removed. Such phosphorus-free sea water has an alkalinity of 2.35 moles/m^3 and a total dissolved carbon content of 2.15 moles/m^3. The carbonate ion content in this water must be equal to 2.35 (A) minus 2.15 (ΣCO_2), or .20 (CO_3^{--}) moles/m^3. The rest of the carbon, 1.95 moles/m^3, must be in the form of bicarbonate ions, since the dissolved, gaseous CO_2 we are omitting from our discussion constitutes only .01 moles/m^3.

Now let us assume that our sample of surface water is transferred to the deep Pacific. Along its way, it receives the oxidation products of enough organic tissue and the dissolution products of enough $CaCO_3$ to raise its total dissolved carbon by about 15 percent. So instead of having a ΣCO_2 content of 2.15 moles/m^3, we have a ΣCO_2 content of 2.45 moles/m^3. An extra amount of carbon, designated $\triangle \Sigma CO_2$, equal to .30 moles/m^3 has been added to this water. If the organic tissue and the $CaCO_3$ are destroyed in roughly the same proportions in which they are formed (so that for every increase of 3 moles in ΣCO_2 there is an increase of 2 moles in A), then the alkalinity increase is .20 moles/m^3. Since surface water has an alkalinity of 2.35, our Pacific deep water must have an alkalinity of 2.55. Although Pacific Deep Water has a higher alkalinity and a higher total dissolved carbon content than cold surface water, it has a lower carbonate ion content. The difference between A and ΣCO_2 in our deep water sample is only 2.55 minus 2.45, or .10 moles/m^3. Despite the fact that Pacific Deep Water has 15 percent more total dissolved carbon, it has only one-half the carbonate ion content of cold surface water because more CO_2 is released to deep water by the oxidation of organic tissue than CO_3^{--} ion is released by the solution of $CaCO_3$. The excess carbon dioxide combines with carbonate ion to form bicarbonate ($CO_2 + CO_3^{--} + H_2O \rightarrow 2HCO_3^-$). The HCO_3^- ion content of the Pacific Deep Water is 2.35 moles/m^3, which is, of course, higher than that in surface water. Table 2-5 summarizes these calculations, and Figure 2-4 illustrates a convenient way to remember the effects of ΣCO_2 and A changes.

In Chapter 1, we discussed the factors that determine the distribution of various elements in the sea. Those same factors also determine the carbonate ion variation in the deep sea. As we go from the North Atlantic to the Pacific, a progressive increase in carbon content of 15 percent occurs (Figure 2-3). The corresponding alkalinity increase is about 8 percent. The CO_3^{--} ion content drops from .20 in newly formed NADW to .10 in Pacific Deep Water. The factors that *enrich* the nutri-

Table 2-5   Carbonate chemistry of various water types.

Water Type	Gaseous CO_2, moles/m³	Bicarbonate Ion (HCO_3^-), moles/m³	Carbonate Ion (CO_3^{--}), moles/m³	Total Dissolved Carbon ($CO_2 + HCO_3^- + CO_3^{--}$), moles/m³	Alkalinity ($HCO_3^- + 2CO_3^{--}$), moles/m³
Warm Surface	.01	1.65	.35	2.01	2.35
Cold Surface	.01	1.95	.20	2.16	2.35
Deep Atlantic	.015	2.10	.15	2.26	2.40
Deep Pacific	.02	2.35	.10	2.47	2.55

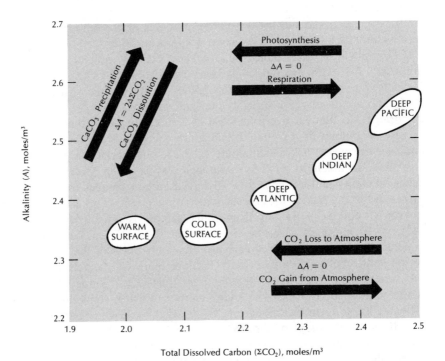

Figure 2-4   Relationship between the total dissolved inorganic carbon content ΣCO_2 and the alkalinity A of waters from various parts of the ocean (a so-called "Deffeyes diagram"). Arrows indicate the effects of various processes occurring within the sea. Of these, only the formation and the destruction of $CaCO_3$ cause the alkalinity to vary. The warm–cold surface ocean difference is due to the transfer of gaseous CO_2 to and from the atmosphere. The cold surface–deep water sequence is due to a combination of respiration and $CaCO_3$ dissolution.

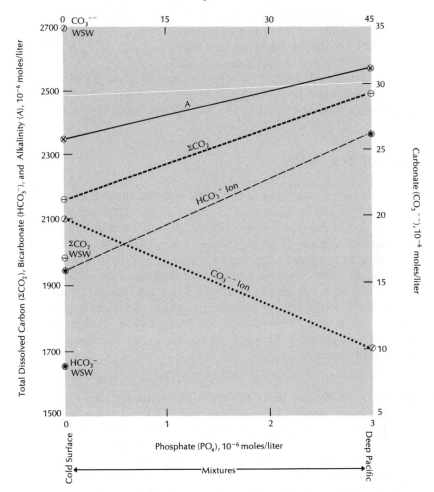

Figure 2-5 Carbonate (CO_3^{--}) ion content, bicarbonate (HCO_3^-) ion content, total dissolved inorganic carbon (ΣCO_2) content, and alkalinity A of various water masses as a function of their critical nutrient content for an ideal, borate-free ocean. The lines join the two end members, cold surface water and Pacific Deep Water. The values for warm surface water (WSW) are shown as discrete points on the left-hand axis.

ents in the most remote parts of the deep ocean correspondingly *reduce* the CO_3^{--} ion content (see Figure 2-5).

Warm and *cold* surface ocean water have been carefully distinguished in discussions thus far. As we will see in Chapter 4, gaseous CO_2 is transferred through the atmosphere from the warm to the cold surface ocean. This transfer leads to a .15 mole/m³ depletion of carbon in tropical water relative to polar water. The transfer of carbon does not alter the alkalinity, however; like cold surface water, warm surface water has an alkalinity of 2.35 moles/m³. But unlike cold water, the ΣCO_2 content of warm surface water is only 2.00 moles/m³. The CO_3^{--} ion content of warm surface water is thus .35 mole/m³, 3.5 times higher than that of Pacific Deep Water!

In summary, the CO_3^{--} ion content ranges from .35 mole/m³ in warm surface water to .10 mole/m³ in the deep Pacific. Its distribution pattern is the inverse of that for the biologically active elements.

The Acidity of Sea Water

Before we take the last step in our effort to relate the degree of saturation of calcite and aragonite in sea water to the distribution of these minerals in sediments, let us sidetrack briefly to consider the pH (the hydrogen ion content or acidity) of sea water. Actually, our approach permits us to completely eliminate the consideration of hydrogen ions. However, as natural waters are often characterized by their pH, we will mention here how acidity varies within the sea. The pH of sea water is determined by the ratio of bicarbonate ions to carbonate ions because that ratio determines the hydrogen ion concentration. To see why this is so, we must consider the following chemical reaction among the ions dissolved in the sea:

$$HCO_3^- \rightleftharpoons CO_3^{--} + H^+$$

A bicarbonate ion can break up into a carbonate ion plus a hydrogen ion, or a carbonate ion can combine with a hydrogen ion to make a bicarbonate ion. As this reaction takes place with great speed, sea water can be assumed to have an equilibrium mixture of these three ions. As in any chemical reaction, the product of the concentration of the ions on one side divided by the product of the concentrations of the ions on the other side is a constant at any given temperature and pressure:

$$k = \frac{[CO_3^{--}]\,[H^+]}{[HCO_3^-]}$$

or

$$[H^+] = k\,\frac{[HCO_3^-]}{[CO_3^{--}]}$$

where k is the constant.

So, if we know the ratio of bicarbonate ions to carbonate ions at a given point in the sea and if we know the temperature and pressure (which define the value of the equilibrium constant k), we can calculate the hydrogen ion content. Let us compare Pacific Deep Water and Atlantic Deep Water (at a level of 5000 meters, the values of their k's would be nearly the same since the temperatures and the pressures are the same). The ratio of the hydrogen ion concentration in deep Pacific water relative to deep Atlantic water would be given by the ratio of the bicarbonate ion in the deep Pacific to that in the deep Atlantic multiplied by the ratio of the carbonate ion in the deep Atlantic to that in the deep Pacific:

$$\frac{[H^+]_{DP}}{[H^+]_{DA}} = \left(\frac{[HCO_3^-]_{DP}}{[HCO_3^-]_{DA}}\right)\left(\frac{[CO_3^{--}]_{DA}}{[CO_3^{--}]_{DP}}\right)$$

Using the values in Table 2-5, we obtain:

$$\frac{[H^+]_{DP}}{[H^+]_{DA}} = \left(\frac{2.45 - .10}{2.25 - .20}\right)\left(\frac{.20}{.10}\right) = 2.4$$

So the H^+ ion content of deep Pacific water is 2.4 times higher than it is in deep Atlantic water. This means that Pacific Deep Water is more acidic (that is, it has a lower pH). The additional acid is the CO_2 (carbonic acid) derived from the oxidation of organic debris. Since more of this debris has been oxidized in the deep Pacific than in the deep Atlantic, Pacific Deep Water has a higher acidity.*

You may wonder why we did not include the hydrogen ion in our calculations of the charge balance in the preceding section. The reason is that the H^+ ion content in sea water is about 10^{-5} moles/m³ (several orders of magnitude below that of the other ions with which we have been concerned). Its contribution to the charge balance is thus negligible. This is also true for the hydroxyl ion content [OH⁻], which is about 10^{-3} moles/m³.

Location of the Saturation Horizon

With the combined knowledge of the variations in the saturation concentration of CO_3^{--} ions in sea water with temperature and pressure and of the distribution of CO_3^{--} ions in the sea, we can determine the degree of saturation throughout the sea. Figure 2-6 plots calculations of the degree of saturation D for the minerals aragonite and calcite as a function of depth in the Atlantic and Pacific oceans. The surface water of the ocean is highly supersaturated with respect to both calcite and aragonite. This degree of saturation drops off very rapidly through the main thermocline (100–1000 meters), producing a crossover to undersaturation for aragonite in the range of 200–400 meters in the Pacific. This is consistent with the absence of aragonitic pteropods in deep Pacific sediments; in any part of the Pacific where water depth is more

---

* Although part of this added CO_2 is neutralized by dissolving $CaCO_3$, the neutralization is not complete, as mentioned above. The remaining CO_2 raises the deep water acidity.

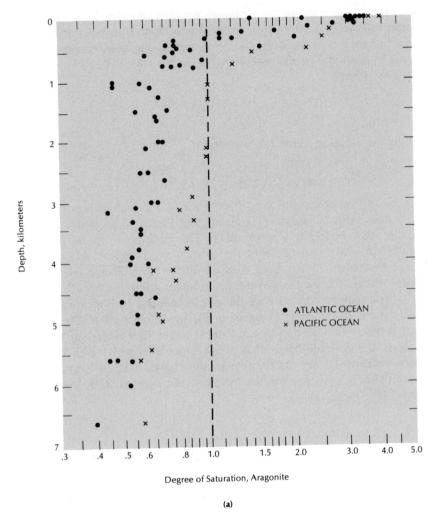

Figure 2-6 The degree of supersaturation as a function of depth in the Atlantic and Pacific Oceans for aragonite (a) and for calcite (b). (Data collected by Yuan-Hui Li, Lamont-Doherty Geological Observatory.)

than a few hundred meters, pteropods fall into undersaturated water and dissolve. In the North Atlantic, the water becomes undersaturated with respect to aragonite at about 2000 meters. The fact that pteropods are found in Atlantic sediments to a depth of 2500 meters is consistent with these measurements. The difference in the degree of saturation between the two oceans is due, of course, to the fact that the deep Pacific has a lower carbonate ion content at any given depth than the deep Atlantic (the temperatures and pressures are not significantly different).

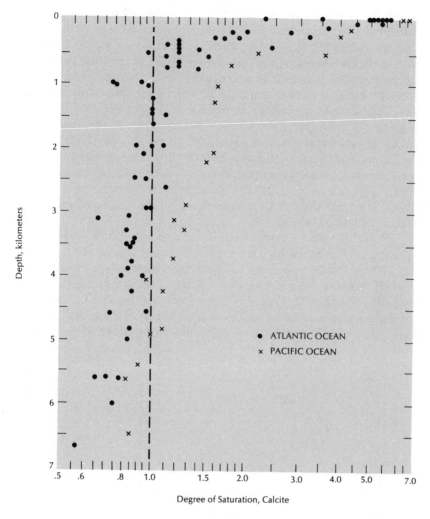

(b)

Calcite is less soluble than aragonite, so in each ocean the water remains supersaturated with respect to calcite to much greater depths than with respect to aragonite. In the Atlantic, the crossover point for calcite is at about 4500 meters, which is consistent with the depth at which calcite begins to disappear from the sediments. In the Pacific, the crossover for calcite is somewhere between 400 and 3500 meters. The precise point is difficult to determine because of the near constancy of D over this depth range and the experimental uncertainties involved (± 10 percent in the calculated value of D). It is, however, consistent with a shallower onset of the disappearance of calcite from the Pacific than from the Atlantic.

Now we are in a position to see what controls the $CaCO_3$ content of sediments in the ocean, and why roughly only 20 percent of the

CaCO₃ is preserved in the world ocean while about 80 percent is destroyed. If we measured the area of only those regions of the ocean floor that projected above the saturation horizon, we would find it to be roughly 20 percent. About 80 percent of the ocean floor lies below this horizon and is bathed in water that is undersaturated with respect to calcite (and, of course, aragonite as well). Over this entire area, red clay (alumino-silicate debris) is raining at an average rate of about .3 g/cm² 10³ yrs (grams per square centimeter per 1000 years), as is CaCO₃ at an average rate of about 1.0 g/cm² 10³ yrs. If the raining CaCO₃ lands above the saturation horizon, it is preserved and the sediments consist of 3 parts clay and 10 parts CaCO₃. (These sediments are called *carbonate oozes*.) Below the saturation horizon, most of the CaCO₃ falling to the sea floor redissolves and the sediments are dominated by alumino-silicates.

Along the East Pacific Rise, a traverse at 17° S reveals that sediments on top of the Rise are about 85 percent CaCO₃. They remain very high in CaCO₃ to a depth of about 3950 meters (see Figure 2-7), at which point a sudden transition occurs to sediment containing 15 percent or less CaCO₃. Assuming the accumulation rate of the non-carbonate component of the sediment and the rate of calcite fall are the same across this boundary, then more than 97 percent of the calcite

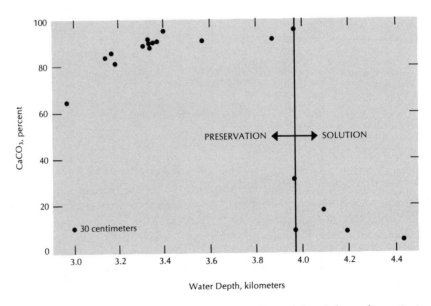

Figure 2-7 Content of CaCO₃ (in percent by weight of dry sediment) at a depth of 30 centimeters in the sediment column as a function of bottom depth for a series of cores collected along a traverse extending west from the crest of the East Pacific Rise at 17° S latitude.

must undergo dissolution below the transitional horizon. (Above the boundary, for every 15 units of noncarbonate 85 units of carbonate accumulate; below the boundary, for every 15 units of noncarbonate no more than 2 units of calcite survive.)

The upward slope in the saturation horizon from the Atlantic to the Pacific combined with the greater average depth of the Pacific give rise to a rather important difference in sediment type between the two oceans. Roughly 60 percent of the Atlantic bottom and only 15 percent of the Pacific bottom lie in waters which permit $CaCO_3$ to accumulate. Red clay thus dominates the Pacific; calcite ooze, the Atlantic. We have already pointed out that the very factor which causes the tilt of the saturation horizon (the deep current from Atlantic to Pacific) also deprives the Atlantic sediments of opal. Thus the general pattern of deep mixing is strongly recorded in the sediment by both opal and calcite distribution.

One question that arises in this connection is why, like opal, is the production of $CaCO_3$ not suppressed in the Atlantic. This suppression would tend to counterbalance the tendency for a higher $CaCO_3$ content in Atlantic than in Pacific sediments. The answer lies in the fact that nitrate and phosphate are not nearly as efficiently driven to the Pacific deeps as is silica. As we saw in Chapter 1, there are two reasons for this: N and P are efficiently recycled within the surface ocean while Si is not, and N- and P-bearing particles are destroyed on the average at much shallower depths than are Si-bearing particles. These two reasons are, of course, due to the same phenomenon: organic tissue fuels animals; opal does not. Thus, whereas the waters upwelling in the Atlantic are silica starved, nitrate and phosphate are adequate to support a $CaCO_3$ productivity rivaling that in the Pacific.

Variation of Sediment Type with Time

A map of the $CaCO_3$ distribution in sediments of another age could be expected to be strikingly different from such a map for the present-day ocean. One reason is that the depth and tilt of the saturation horizon could have varied with time (we will consider this possibility in Chapter 7). For now, if we assume that the level of $CaCO_3$ saturation has remained roughly constant with time over the last few tens of millions of years, then what composition changes would be expected with depth in the sedimentary column at any point on the sea floor? Geophysicists have shown that the ocean crust is moving away from the crests of the mid-ocean ridges at the rate of a few centimeters a year. New crust is constantly being generated to fill the void this movement creates (see Figure 2-8). As the newly formed crust or *lithospheric plate* moves

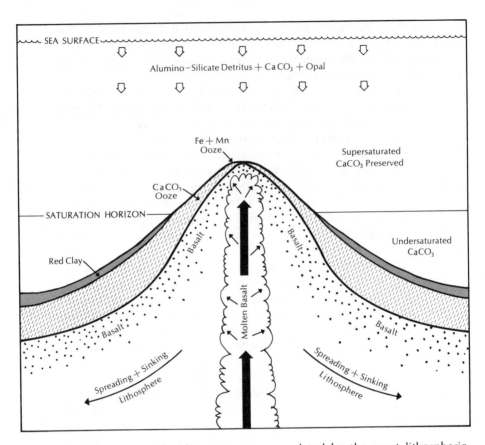

Figure 2-8 Sequence of sediment types accumulated by the great lithospheric plates as they move away from the crests of mid-ocean rises. The first sediment to be deposited is iron (Fe) + manganese (Mn) oxide, a product of volcanism. When a point a few kilometers away from the crest is reached, the sediment no longer receives volcanic products and is dominated by $CaCO_3$ falling from the surface. At a point several hundred kilometers from the crest, the plate subsides below the saturation horizon for calcite and $CaCO_3$ no longer accumulates. Beyond this point, continental detritus and perhaps opal dominate the sediment. A core bored through sediment capped with red clay would encounter buried $CaCO_3$ and then a thin layer of Fe + Mn-rich sediment before entering the underlying basalt (hard rock).

away from the crests, it accumulates more and more sediment. Indeed, the thickest sediments are found furthest away from the mid-ocean ridges. As the ridge crests are approached, the sediment thins down to nothing.

In addition to this lateral motion, there is a vertical motion. The mid-ocean ridge crests have a more or less constant elevation, projecting to within 3000 meters of the ocean surface. The ocean floor is well away from the ridge crests, 6000 meters below the sea surface (actually, by this point, 500 meters of sediment have accumulated and the water depth is only 5500 meters). As the ocean floor moves away from the ridge crests, the crests lose 3000 meters in elevation. Viewing this whole process, we see that at the time any particular portion of crust was first formed it was bathed in $CaCO_3$ supersaturated water. However, as the crust moved away from the ridge, it eventually dropped below the saturation horizon. At this point, $CaCO_3$ accumulation should have ceased. The $CaCO_3$ already accumulated would be gradually buried beneath a layer of red clay (Figure 2-8). Over the rest of the trip across the sea floor, only alumino-silicate debris would accumulate. So we would expect that if a hole were drilled down through the sediments in any region where the sea floor currently lies below the saturation horizon, the clay encountered at the top of the core would eventually give way to carbonate-rich sediment. This sequence has indeed been found in numerous deep sea drillings.

The movement of the ocean floor will often carry a segment of the crust across the equator (see Figure 2-9). Since opal deposition today

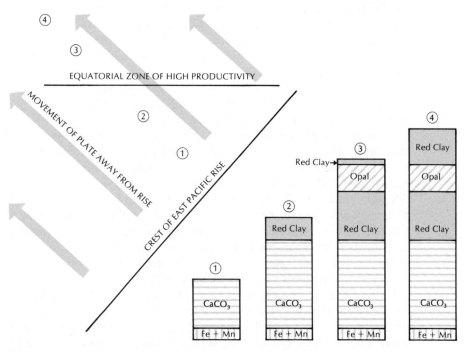

Figure 2-9   Sequence of sediments accumulating on a plate which crosses the high productivity equatorial belt in its movement away from a ridge crest. The numbers relate the map locations to the corresponding sedimentary sequences found in borings made at these points.

is very rapid along the equator in the Pacific Ocean, we would expect to find a silica-rich horizon in the sedimentary record for any crustal segment which made such a crossing. Only during the period when the crust lay beneath the high productivity zone, would it accumulate Si-rich sediment; the overlying and underlying sediments would be silica poor by comparison. Examples of such a sequence uncovered by deep drilling are shown in Figure 2-9.

Summary

Ocean sediments are dominated by two major components:* alumino-silicate debris (red clay) from the continents and SiO_2 and $CaCO_3$ hard parts formed by marine organisms. The clay debris decreases in abundance away from the continental margins. Opal accumulates most rapidly beneath those areas of the ocean where enhanced upwelling of Si-rich water occurs. As all ocean water is undersaturated with respect to opal, dissolution attack occurs everywhere. By contrast, $CaCO_3$ is produced more uniformly over the ocean surface, but it falls into water which is in some places supersaturated and in others undersaturated. In any given ocean basin, $CaCO_3$ is absent in the deepest sediments but present on the crests of ridges and islands which project above the sea floor.

The distribution of both opal and calcite in sediments strongly reflects the horizontal segregation of biologically utilized elements in the deep sea. Opal is rare in Atlantic sediments as a consequence of the marked deficiency of silica in that ocean. Calcium carbonate is less abundant in Pacific than in Atlantic sediments because of the difference in carbonate ion content between the deep waters of the two oceans. The waters of the Pacific are capable of destroying carbonate at considerably shallower depths than Atlantic waters. Thus the circulation patterns which lead to chemical segregations in the sea are strongly reflected by the composition of the sediments. Because of the horizontal movement of the ocean crust away from the mid-ocean ridge crests and to greater depths, the temporal sequence of sediment types on any given portion of the crust will vary in a predictable way.

In this chapter, we have again painted a somewhat idealized picture for reasons of clarity. The details of opal and calcite destruction are just starting to emerge. When they have all been assembled, our ideas will surely change. Although most of these changes are expected to be in details, as in Chapter 1, there is a chance that the main message presented here will be significantly altered.

---

* The third component, authigenic minerals, will be discussed in Chapter 4.

■ Publications dealing with the distribution of sediment components on the sea floor:

Lisitzin, A. P. *Sedimentation in the World Ocean.* Tulsa, Oklahoma: Society of Economic Paleontologists and Mineralogists Special Publication No. 17, 1972. A broad treatment of the distribution of sediment types throughout the world ocean.

National Science Foundation, National Ocean Sediment Coring Program. *Initial Reports of the Deep Sea Drilling Project.* Washington, D.C.: U.S. Government Printing Office. Complete lithologic and paleontologic descriptions of borings made throughout the deep sea by the drilling ship *Glomar Challenger.*

Biscaye, P. E. "Mineralogy and Sedimentation of Recent Deep-Sea Clay in the Atlantic Ocean and Adjacent Seas and Oceans." *Bull. Geol. Soc. Amer.* 76 (1965):803–32. The distribution of various clay minerals in recent sediments of the Atlantic Ocean basin.

Emery, K. O. *The Sea Off Southern California.* New York: John Wiley & Sons, Inc., 1960. A discussion of sedimentary processes occurring along a continental margin.

■ Publications dealing with $CaCO_3$ and SiO_2 dissolution on the sea floor:

Berner, Robert A. *Principles of Chemical Sedimentology.* International Series in the Earth and Planetary Sciences. New York: McGraw-Hill, 1971. A textbook dealing with the processes that produce chemical alteration in sediments.

Berger, W. H. "Planktonic Foraminifera: Selective Solution and the Lysocline." *Marine Geology* 8(1970):111–38. A discussion of the productivity and dissolution patterns of foram shells in the world ocean.

Peterson, M. N. A. "Calcite: Rate of Dissolution in a Vertical Profile in the Central Pacific." *Science* 154(1966):1542–44. A clever experiment aimed at the direct determination of the rate of dissolution of calcite as a function of depth in the Pacific Ocean.

Berner, Robert A., and Wilde, P. "Dissolution Kinetics of Calcium Carbonate in Sea Water: I. Saturation State Parameters for Kinetic Calculations." *Amer. Jour. Sci.* 272(1972):826–39. A laboratory study of the variation of the rate of calcite dissolution with the degree of undersaturation of sea water.

Hurd, David C. "Interactions of Biogenic Opal, Sediment, and Seawater in the Central Equatorial Pacific. *Geochim. Cosmochim. Acta* 37(1973): 2257–82. A laboratory study of the factors controlling the rate of opal dissolution.

Broecker, Wallace S. "Calcite Accumulation Rates and Glacial to Interglacial Changes in Oceanic Mixing." *The Late Cenozoic Glacial Ages,* ed.

K. K. Turekian. New Haven, Conn.: Yale University Press, 1971, pp. 239–65. Evidence for glacial to interglacial changes in the rate of $CaCO_3$ production and in the degree of dissolution of this $CaCO_3$.

Broecker, Wallace S., and Broecker, S. "Carbonate Dissolution on the Western Flank of the East Pacific Rise." *Studies in Paleo-Oceanography,* ed. W. W. Hay. Tulsa, Oklahoma: Society of Economic Paleontologists and Mineralogists Special Publication 20, 1974. A detailed study of the nature of the transition from $CaCO_3$-rich to $CaCO_3$-poor sediments on the west flank of the East Pacific Rise.

■ Publications dealing with plate tectonic motions:

Geology Today. Del Mar, California: CRM Books, 1973. General geology textbook with good coverage of this subject.

Continents Adrift: Readings from Scientific American. Introduction, J. Tuzo Wilson. San Francisco: W. H. Freeman & Company, 1972. A compilation of *Scientific American* articles on this subject.

Gass, I. G. et al. *Understanding the Earth: A Reader in the Earth Sciences.* Cambridge, Mass.: The M.I.T. Press, 1971. A textbook containing an in-depth treatment of the current mode of plate motion and the sequence of events leading to its discovery.

PROBLEMS

2-1   If Pacific Deep Water were returned to the laboratory and stored under the following conditions, how would its calcium ion content, carbonate ion content, and pH change:

    (a) at $2°$ C, 1 atm pressure?
    (b) at $2°$ C, 1 atm pressure with powdered calcite?
    (c) at $2°$ C, 1 atm pressure with powdered aragonite?
    (d) at $24°$ C, 1 atm pressure?

Assume that precipitation of $CaCO_3$ occurs only if a mineral substrate is present and that the mineral formed will be the same as the substrate.

2-2   What is the alkalinity of each of the following solutions?

    (a) One mole of NaCl dissolved in 400 liters of CO_2-free distilled water.
    (b) One mole of Na_2CO_3 dissolved in 1000 liters of CO_2-free distilled water.
    (c) One mole of $CaSO_4$ dissolved in 5000 liters of CO_2-free distilled water.
    (d) One mole of $CaSO_4$ and 2 moles of $Ca(HCO_3)_2$ dissolved in 3000 liters of CO_2-free distilled water.

2-3   Organic tissue containing 2 moles of carbon is added to a 1000-liter sample of Pacific Deep Water. The bacteria in the sample respond and "burn" the organic matter in the residual oxygen (.05 moles/m³). Which respiration ingredient will run out first—oxygen or tissue? At this ingredient's point of disappearance, by how much will the carbonate ion content have changed? What might happen if this water were stored longer?

2-4   In a given area, the upwelling rate of Pacific Deep Water is 10 m³/yr for each square meter of sea surface. If the area in question is shallow enough so that the bottom lies in calcite supersaturated water, what would be the rate of sedimentation of calcite in g/cm² 10³ yrs?

2-5   If the rate of sea floor subsidence averages 3 meters per million years, what is the age of the sea floor when it passes through the level of $CaCO_3$ dissolution in the Atlantic Ocean? (Assume the average ridge crest elevation to be 3000 meters and the dissolution horizon to be at 4500 meters.) If the mean rate of accumulation of noncarbonate sediment is .5 cm/10³ yrs, how deep would you have to bore into the sea floor to reach $CaCO_3$-bearing sediment at a depth of 5000 meters in the Atlantic? If the mean $CaCO_3$ accumulation rate at depths shallower than the dissolution horizon is 1.5 cm/10³ yrs, how much further would you have to bore to reach the basalt at the base of the $CaCO_3$-bearing sediment?

3

HOW FAST DOES THE MILL GRIND?

In Chapter 1, we stated that the residence time of water in the deep sea is 1600 years. In Chapter 2, the rate of rain of CaCO$_3$ toward the deep sea floor was reported to be 1 g/cm^2 10^3 yrs. How were these rates established? Neither the pattern of the inhomogeneous composition of sea salt nor of the major sedimentary components tells us anything about the rates at which the system runs. As we will see in this chapter, these rates are based primarily on the measurement of trace quantities of a radioisotope of carbon which travels through the sea and into the sediments along with the components of sea salt we have already discussed. Let us first see how the distribution of this isotope can be made to divulge the residence time of water in the deep sea.

Rate of Vertical Mixing

A radioisotope of any given element has a chemistry identical to that of the element's nonradioactive or "normal" isotope. So a radioactive carbon atom moves through the ocean just as any other carbon atom does. Since we have already worked out the cycle of carbon in sea water, we can use this information in connection with the application of carbon-14 (C-14) as a timekeeper.

Radioactive carbon is continually being produced in our atmosphere. There, cosmic rays from interstellar space encounter atoms within the atmosphere and fragment some of them, releasing neutrons from their nuclei. Many of these neutrons find their way to nitrogen atoms (nitrogen gas makes up 80 percent of our atmosphere). A neutron entering a nitrogen-14 (N-14) nucleus can knock out a proton and remain behind in its place (like a cueball in pool). Now, as the nucleus of a N atom has 7 protons and the nucleus of a C atom has 6 protons, the act of knocking one proton out of the nitrogen nucleus changes it into a carbon nucleus. The number of nucleons (that is, the number of neutrons plus protons) in the new atom is still 14, the same number as in the old atom. The difference is that one neutron has entered the nucleus and one proton has been expelled. The C-14 nuclei produced are unstable; the mixture of 8 neutrons and 6 protons is not a durable one, and the C-14 nucleus endeavors to change itself back into the stable 7–7 arrangement of N-14. To do so, the C-14 nucleus must hurl out an electron. This converts one neutron into a proton and the nucleus again assumes the form of N-14. (The cycle is shown in Figure 3-1.)

About 100 C-14 atoms are being generated by cosmic rays over each square centimeter of the earth's surface each minute. Over many thousands of years, the amount of C-14 on the surface of the earth has become constant.* It is disappearing by its own radioactive decay at the same rate that it is being produced by cosmic rays (100 atoms of C-14 undergo radioactive decay per square centimeter of earth surface each minute).

The property that makes C-14 a very interesting and important isotope in nature is that the average C-14 atom requires 8200 years to undergo the transformation back to N-14! Thus, in their lifetimes, C-14 atoms have a chance to mix reasonably well *but not completely* with the ordinary carbon in the atmosphere and in the ocean. The fact that the mixing is *not* quite complete makes C-14 a very valuable tool in oceanography.

Before exploring the tracer potentials of C-14, let us first review the rules governing radioactive decay. The disappearance of a batch of radioactive atoms is a logarithmic process: a set fraction of the atoms undergoes radioactive transformation each year. For C-14, this fraction is 1 atom in 8200 each year. The average C-14 atom lives 8200 years. This logarithmic decay pattern gives rise to the concept of

---

* The concentration of a radioisotope is often expressed in radioactivity rather than mass units. Since the mass of a given radioactive species is proportional to the rate at which it undergoes radioactive decay, stating the rate of radioactive decay of an isotope in a given volume of water is also stating its mass. The units of radioactivity are disintegrations per unit of time: disintegrations per second (dps), disintegrations per minute (dpm), and so on. Since radioisotopes are generally measured by intercepting the radiations emitted during the disintegration of an atom, it proves more convenient to use activity rather than mass units.

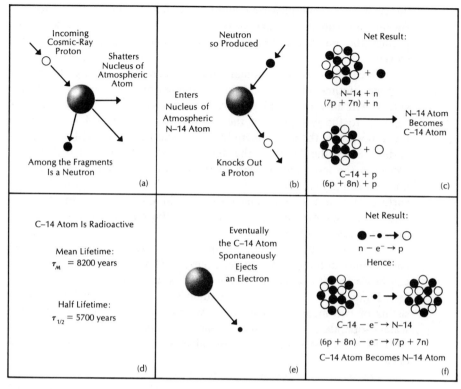

Figure 3-1 The "life cycle" of a carbon-14 atom. Created in the atmosphere by the collision of a neutron (produced by primary cosmic-ray protons) with a nitrogen atom, the average C-14 atom "lives" for 8200 years. Its life is terminated by the ejection of an electron which returns the atom to its original form, N-14.

half-life. If we were to start with a given number of radiocarbon atoms today and let them age for 5700 years, half of them would have become N-14 atoms and the other half would still be C-14 atoms. Half of the half of C-14 atoms that remained would undergo transformation in the next 5700 years, and so forth.

For an analogous situation, consider a roomful of students. Each has a coin which he flips once every second. No one can leave the room until he has flipped ten heads in a row. One student might be lucky enough to accomplish this feat in the first few minutes; another might not have succeeded even after several hours of steady flipping. From the laws of probability, however, we could predict exactly what the average success time would be. The fraction of students present who would leave in a given time interval would be the reciprocal of this average success time. If there were enough students in the room when the flipping began, we would observe that the number remaining would decrease in the same halving pattern that holds for the radioactive decay of C-14. Although we could not predict the time required for any specific individual to make a successful run of ten heads, we could predict the number of departures for a known time interval rather closely.

Since there is not enough time for complete mixing during the average lifetime of the C-14 atom, the more "remote" parts of the system will have less C-14 per unit of ordinary carbon than those parts "closer" to the birthplace of the C-14 atoms. Since C-14 is created in the atmosphere, the atmosphere has the highest ratio of C-14 to ordinary carbon. Since the "end" of the deep current in the deep North Pacific Ocean is the most remote part of the ocean, it has the least C-14 per unit of ordinary carbon. The C-14/C ratios that interest us are those for surface and deep water. The difference in C-14/C contents between these two reservoirs allows us to establish the rate of mixing across the main thermocline of the ocean.

If we want to make our estimates rigorous, we need a model that corresponds to the ocean in its complexity. However, the first principles can be demonstrated by the simple two-box model we used in Chapter 1 (see Figure 3-2). If we can understand how a hypothetical ocean consisting of two major water masses—warm surface water and cold deep water—operates, we will have a basis for understanding the methods used to solve the vastly more complicated problem involving the real ocean.

For the deep reservoir of our two-box ocean, three substances must be conserved: water, ordinary carbon, and C-14. We assume that the surface and deep reservoir sizes and their carbon contents have achieved constant values. These reservoir sizes and the distribution of ordinary carbon between them in no way betrays the rate of mixing. To find the rate we need our radioactive clock, C-14. Because C-14 is disappearing by radioactive decay, its distribution between the surface and deep reservoirs *does* depend on the rate of mixing!

Thus we will write three equations: one for water, one for ordinary carbon, and one for C-14. The water conservation equation very simply states that the amount of water going up v_{up} must equal the amount of water coming down v_{down}. We set both v_{up} and v_{down} equal to the parameter v_{mix}:

$$v_{up} = v_{down} = v_{mix}$$

As stated in Chapter 1, v_{mix} is the rate at which water is exchanged between the two reservoirs.

Carbon is added to the deep reservoir in two ways: it comes down with the descending surface water and it falls from the surface in the form of particles destined for destruction in the deep sea. The sum of these two contributions must be exactly balanced by the amount of carbon carried away by upwelling water. We can write the balance between these fluxes in the form of the equation:

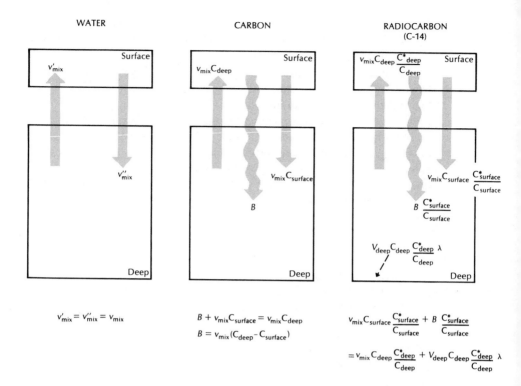

Figure 3-2 Two-box model for the cycle of water, carbon, and radiocarbon (C-14) between the surface and deep sea. If the system is at steady state, then the amount of each of these substances entering the deep reservoir must exactly match the amount lost. Although fluxes of carbon and C-14 from rivers and to the sediments should be included, they are so small compared to the other fluxes that they are negligible. The solid arrows represent the fluxes of substances carried by water; the wavy arrows, substances carried by particles. The dashed arrow represents loss by radioactive decay.

$$v_{mix}C_{deep} = v_{mix}C_{surface} + B$$

where C_{deep} is the carbon content of upwelling deep water, $C_{surface}$ is the carbon content of descending surface water, and B is the carbon added to the deep reservoir each year by the destruction of particles falling from the surface. Solving for B, we obtain:

$$B = v_{mix}(C_{deep} - C_{surface})$$

A similar equation can be established for C-14. Again, we have two water fluxes. First, C-14 is carried up from the deep to the surface water and down from the surface to the deep water as the result of physical mixing. Second, C-14 enters the deep water with descending particles. But there is still another factor to consider here: one part in

8200 of all the radiocarbon in the deep ocean undergoes radioactive decay and disappears each year. We therefore need a term that will give us the radioactive decay rate.

Radiocarbon is measured in sea water samples by extracting carbon from the water and measuring that carbon's radioactivity. This shows the amount of C-14 contained in the dissolved carbon, or the C-14/C ratio.* Laboratory measurements thus do not give the absolute amount of radiocarbon, but the ratio of radiocarbon atoms (designated C^*_{deep} or $C^*_{surface}$) to ordinary carbon atoms (C_{deep} or $C_{surface}$). So the amount of C-14 in descending surface water is equal to the volume of descending water v_{mix} times the carbon content of this descending water $C_{surface}$ times the ratio of C-14 atoms to ordinary C atoms in this carbon:

$$v_{mix} C_{surface} \left(\frac{C^*_{surface}}{C_{surface}} \right)$$

The amount of C-14 in upwelling deep water is equal to v_{mix} times the concentration of carbon in deep water C_{deep} times the ratio of C-14 atoms to ordinary C atoms in deep water:

$$v_{mix} C_{deep} \left(\frac{C^*_{deep}}{C_{deep}} \right)$$

Particles carry C-14 down in almost exactly the same proportion to other carbon that is present in surface water. So the particle flux will be the amount of carbon descending each year in particles B times the ratio of C-14 atoms to ordinary C atoms in surface water:

$$B \frac{C^*_{surface}}{C_{surface}}$$

To calculate the radioactive decay of C-14, we first find the total number of C-14 atoms in deep water (by taking the volume of deep water V_{deep} times the carbon content of deep water C_{deep} times the ratio of C-14 atoms to ordinary C atoms in deep water) and multiply it by the fraction of C-14 atoms decaying each year, which we designate as λ. Thus:

$$V_{deep} C_{deep} \left(\frac{C^*_{deep}}{C_{deep}} \right) \lambda$$

We can now write the conservation equation for C-14 in the deep reservoir as:

$$v_{mix} C_{surface} \frac{C^*_{surface}}{C_{surface}} + B \frac{C^*_{surface}}{C_{surface}}$$

$$= v_{mix} C_{deep} \frac{C^*_{deep}}{C_{deep}} + V_{deep} C_{deep} \frac{C^*_{deep}}{C_{deep}} \lambda$$

---

* In the ocean, for every C-14 atom there are 10^{12} ordinary carbon atoms. Therefore, C-14 atoms do not comprise the "bulk" of the carbon.

The conservation equation for carbon demanded that the particle flux of carbon B be given by $v_{mix}C_{deep} - v_{mix}C_{surface}$. Thus in the C-14 equation we can eliminate B by substitution. After combining terms, the result is:

$$v_{mix}C_{deep}\frac{C^*_{surface}}{C_{surface}} = (v_{mix}C_{deep} + \lambda V_{deep}C_{deep})\frac{C^*_{deep}}{C_{deep}}$$

Note that each term in this equation contains the concentration of carbon in deep water C_{deep}. Therefore, this common factor can be cancelled. (This is logical, since the rate of vertical mixing cannot be dependent on the amount of carbon dissolved in the sea!) If we then solve the C-14 balance for the unknown mixing rate v_{mix}, we obtain:

$$v_{mix} = \lambda V_{deep}\frac{\dfrac{C^*_{deep}}{C_{deep}}}{\dfrac{C^*_{surface}}{C_{surface}} - \dfrac{C^*_{deep}}{C_{deep}}}$$

We can express the volume of the deep ocean V_{deep} in terms of the product of its mean depth $\hbar$ and its surface area A_{ocean}. The equation then becomes:

$$v_{mix} = \lambda \hbar A_{ocean}\frac{\dfrac{C^*_{deep}}{C_{deep}}}{\dfrac{C^*_{surface}}{C_{surface}} - \dfrac{C^*_{deep}}{C_{deep}}}$$

Dividing numerator and denominator by C^*_{deep}/C_{deep}, we obtain:

$$v_{mix} = \frac{\lambda \hbar A_{ocean}}{\dfrac{C^*_{surface}/C_{surface}}{C^*_{deep}/C_{deep}} - 1}$$

In this equation, λ is equal to $1/8200$ years and $\hbar$ is 3200 meters (3.2×10^5 centimeters). Measurements of the radiocarbon content of carbon extracted from sea water reveal that, as expected, there is a deficiency of C-14 relative to C in deep water with respect to surface water; the C-14 has not completely mixed with the ordinary carbon. The C-14/C ratio in Pacific Deep Water is about 19 percent lower than it is in surface water (see Figure 3-3). When we place these values in the C-14 balance equation and solve, we find that the yearly volume of water exchanged between the surface and the deep ocean is equal in volume to a layer 200 centimeters thick with an area equal to that of the ocean:

$$v_{mix} = \frac{\dfrac{1}{8.2 \times 10^3} \times 3.2 \times 10^5}{1.19 - 1} A_{ocean} = 200 \text{ cm/yr } A_{ocean}$$

Since the mean thickness of the deep reservoir is 3200 meters, the transfer of a layer 2 meters (200 centimeters) thick between the deep

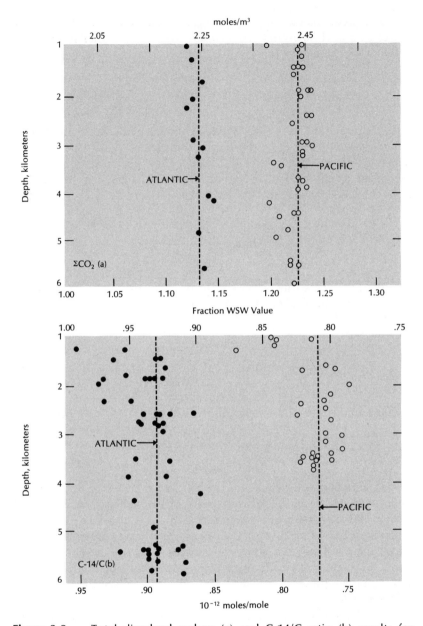

Figure 3-3 Total dissolved carbon (a) and C-14/C ratio (b) results for samples from the deep Atlantic (solid symbols) and from the deep Pacific (open symbols) as a function of depth below 1000 meters. The results are given as fractions of the value for warm surface water as well as in concentration units. Note that ΣCO_2 values increase to the left while those for the C-14/C *decrease* to the left. Pacific Deep Water has a 23 percent higher total dissolved carbon content than warm surface water, but its carbon has a 19 percent lower C-14 content.

and surface ocean each year yields a mean residence time of water in the deep sea of 3200/2, or 1600 years. A unit of water (or salt) has one chance in 1600 of returning to the warm surface reservoir each year.

Rate of Continental Runoff

In Chapter 1 we stated that the ratio of the rate of upwelled water to runoff water entering the warm surface ocean was 20. Now that we have established the upwelling rate, let us turn our attention to the runoff rate. In this case, no radioisotope is needed. A very direct means is used: the amount of water leaving the mouths of the world's rivers is monitored (see Figure 3-4). We can estimate the approximate magnitude of the result of this effort from three pieces of information: the average annual rate of rainfall on the continents (70 cm/yr), the

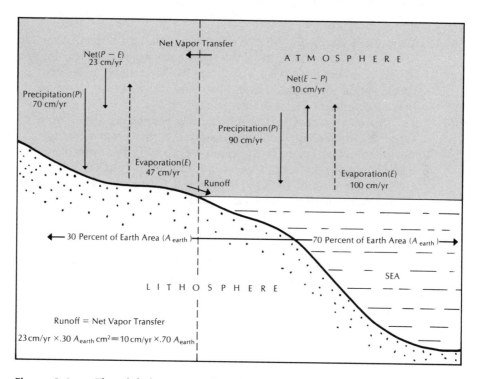

Figure 3-4   The global water cycle. About 10 percent more water evaporates from the ocean than returns to the ocean as rain. The excess falls on the continents as rain and snow. An equal amount of water runs from the continents into the oceans through the world's rivers.

average percent of this water which survives evaporation and reaches the sea (35 percent), and the ratio of the area of the continents to that of the ocean (3 parts continent to 7 parts ocean). With this knowledge, we can calculate that in one year a layer of water 25 centimeters thick with the area of the continents will reach the sea. If spread over the sea, this water would make a layer 10 centimeters thick!

Since the C-14 tracer shows that in a year upwelling brings an amount of water equal to an oceanwide layer 200 centimeters thick, the ratio of v_{mix} to v_{runoff} must be 20:

$$\frac{v_{mix}}{v_{runoff}} = \frac{200 \text{ cm/yr } A_{ocean}}{10 \text{ cm/yr } A_{ocean}} = 20$$

So we see that the value adopted in Chapter 1 is based on both river gauging data and the distribution of C-14 in the ordinary carbon of the ocean.

Predicted Rate of CaCO₃ Rain

We can use the model value of the mixing rate v_{mix} to predict the rate at which $CaCO_3$ is falling from the surface sea to the deep sea. In areas of the deep sea floor lying above the $CaCO_3$ saturation horizon, the accumulation rate should be of the same order of magnitude as the predicted particle fall rate for this substance.

As we discussed in Chapter 1, it is possible to show, from the variations in alkalinity and total dissolved carbon from water mass to water mass, that for every 2 moles of C fixed in the form of organic material roughly 1 mole of C is fixed in the form of $CaCO_3$. The carbon falling in the form of $CaCO_3$ (B_{CaCO_3}) and the total amount of carbon coming down in all particles (B) are therefore in the proportion $1/3$:

$$\frac{B_{CaCO_3}}{B} = \frac{1}{3}$$

We have shown by carbon balance that B must equal the mixing rate v_{mix} times the difference between the carbon content of deep water and surface water:

$$B = v_{mix}(C_{deep} - C_{surface})$$

Hence:

$$B_{CaCO_3} = \frac{1}{3} \, v_{mix}(C_{deep} - C_{surface})$$

We have also shown that v_{mix} is equal to a layer of ocean water 200 centimeters thick per year (or 2 m³/m²). The concentration of C in deep water is 2.45 moles/m³, and in cold surface water it is 2.15

moles/m³. We find, then, that CaCO₃ must be falling from the surface ocean into the deep ocean and dissolving at the rate of $\frac{1}{3} \times 2$ m³/m² yr $\times$ .30 moles/m³, or .20 moles of carbon per square meter of ocean surface per year.

Accumulation rates of material in deep sea sediments are not normally expressed in these units, but rather in units of grams of CaCO₃ per square centimeter of sea floor per 1000 years. So we need to convert from one set of units to the other. It happens that 1 mole of CaCO₃ weighs almost exactly 100 grams. Of course, there are 10^4 square centimeters in a square meter. Thus .20 moles/m² yr of CaCO₃ is equivalent to 2.0 g/cm² 10^3 yrs of CaCO₃, or:

$$B_{CaCO_3} = 2.0 \frac{g}{cm^2\ 10^3\ yrs}$$

Unfortunately, the mixing model tells us only about the CaCO₃ that dissolves and not about the CaCO₃ that falls from the surface and remains in the sediments. If we want to know the total rate of production of CaCO₃, we must add to the amount of CaCO₃ that dissolves the amount that is preserved in the sediments. We can do this by making the assumptions that the CaCO₃ produced above areas of red clay in the ocean is destined to totally redissolve and that the CaCO₃ produced in areas above CaCO₃-rich sediments does not dissolve at all. Since about 20 percent of the ocean floor is covered by CaCO₃-rich sediment, we estimate the total flux of CaCO₃ by assuming that the 2 g/cm² 10^3 years that dissolves is 80 percent of the total; that is, we divide it by .80. This yields a predicted total flux of 2.5 g/cm² 10^3 yrs of CaCO₃. Thus our model predicts that for each square centimeter of ocean surface 2.5 grams of CaCO₃ are being generated every 1000 years. Of this CaCO₃, 2.0 grams fall into undersaturated water and redissolve and the remaining .5 gram accumulates in the sediments. (These calculations are summarized in Figure 3-5.)

Now, if we consider areas of the sea where the ocean bottom projects into supersaturated water, we can measure the rate at which CaCO₃ accumulates and compare that observed rate with our predicted rate. If the rates match, we have an indication that our model is satisfactory; if they fail to match, we must have made some erroneous assumptions.

Sediment Accumulation Rates

Let us turn our attention, then, to the ways in which sedimentation rates are determined. All sedimentation rates in the ocean are based on three primary methods of absolute dating. One involves C-14, the same isotope we used to establish the rate at which the ocean is mixing. A second

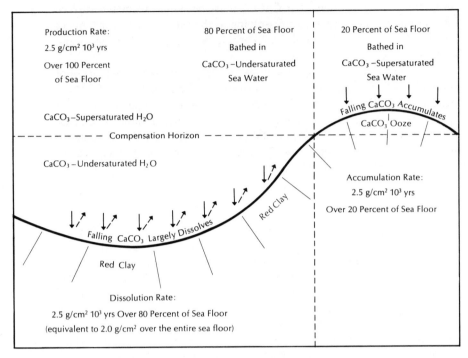

Figure 3-5 Cycle of CaCO$_3$ in the sea. Organisms generate CaCO$_3$ at a fairly uniform rate over the sea surface. About 20 percent of this CaCO$_3$ falls onto portions of the sea floor projecting above the saturation horizon and is preserved in the sediments. The remainder falls into undersaturated water and largely dissolves. The model predicts that for each square centimeter of sea floor 2.5 grams of CaCO$_3$ fall from the surface every 1000 years. Since 80 percent of this dissolves, the deep sea receives an average of 2.0 grams of CaCO$_3$ every 1000 years for each square centimeter of ocean surface. The model also predicts that where the sea floor projects above the saturation horizon, the accumulation rate should be 2.5 g/cm^2 10^3 yrs.

method involves thorium-230 (Th-230) and protactinium-231 (Pa-231), two radioactive isotopes produced by the decay of uranium dissolved within the sea. The third method involves radioactive potassium-40 (K-40), which decays to a gas, argon-40 (Ar-40). These three radio-isotopic methods have all been used to date deep sea sediments and, as we will see, they give nearly the same average accumulation rates.

Deep sea cores contain unique stratigraphic markers which can be traced over large regions of the sea floor. Once a core containing such a marker has been dated, the level in all other deep sea cores where that stratigraphic marker has been found can be assigned the same age. Once the ages of several such horizons have been determined, finding the sedimentation rates is greatly simplified. There are three kinds of marks made in deep sea sediments that can be used for this purpose. One is magnetic. Reversals in the polarity of the earth's magnetic field are recorded in deep sea cores. The times these transitions have been accurately determined by the potassium–argon dating of volcanic rocks. If we can establish which transition is which in a given core, ages can be assigned to the points of polarity reversal. The irregular sequence of these transitions (see Figure 3-6) make this procedure reliable in most cases.

Another series of stratigraphic markers is based on faunal changes. For every group of organisms there are, over periods of millions of years, a few extinctions of old species, a few appearances of new species, and some unique changes in relative abundances. These changes also form key horizons that, when dated, complement the use of magnetic reversals in assigning ages to various levels in a particular deep sea core.

The final stratigraphic marker involves climatic changes. The technique is less reliable because climatic changes are repetitive at intervals of roughly the same duration. In the main, this method is helpful for the last two climate cycles. Further back in time it is too easy to mismatch two curves and end up with the wrong age. However, the last magnetic reversal occurred 700,000 years ago and faunal methods are useful primarily from 400,000 to a few million years ago, so the climatic method's reliability over the last 200,000 years makes

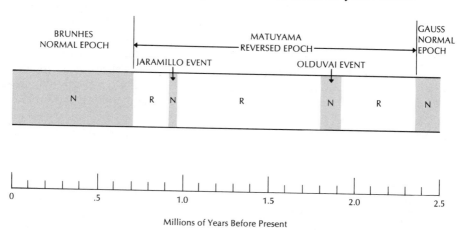

Millions of Years Before Present

Figure 3-6   The time sequence of magnetic reversals as determined by the potassium–argon dating of volcanic rocks. The shaded areas marked *N* (normal) indicate times when the magnetic field had the same polarity it does today, and the open areas marked *R* (reversed) indicate times when it had the opposite polarity. Long intervals of time dominated by either normal or reversed polarity are called *magnetic epochs* and are named after scientists who made important contributions to the field of magnetism. Brief intervals of opposite polarity within these epochs are called *events* and are named after the localities at which they were discovered.

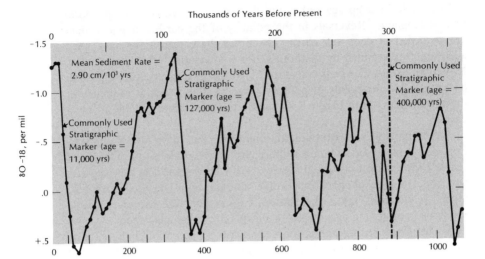

Figure 3-7   Changes in the O-18/O-16 ratio in the CaCO₃ of shells of the plank-tonic species of foraminifera *Sacculerifera* with depth in deep sea core P6304-8, taken by the Miami University research vessel *Pilsbury* in the Caribbean Sea. The lower the O-18/O-16 ratio, the warmer the climate at the time the shell grew. The sharp changes in O-18 content at 40 and 350 centimeter depths in the cores repre-sent the rapid warmings which heralded the end of the last two periods of large scale glaciation. The most recent of these events has been dated by the C-14 method to be 11,000 years in age; the other has been dated by the Th-230 and Pa-231 methods to be 127,000 years in age. These sudden climatic changes imprint them-selves on deep sea cores over a broad expanse of the sea floor, not only through an O-18 change but also in a variety of other ways, and provide very useful stratigraphic markers. δO-18 is the difference (given as parts per thousand) between the O-18/O-16 ratio in the foram sample and the O-18/O-16 ratio in Cesare Emiliani's laboratory standard.

it a valuable tool. (The pattern of climatic change for this period is shown in Figure 3-7.) Climatic changes are recorded in deep sea cores in many ways, including CaCO₃ content changes, O-18/O-16 ratio changes in the CaCO₃ from foram shells, faunal changes, coarse frac-tion changes, and so on.

Radiocarbon Dating

Now let us examine each of the dating methods, beginning with the one based on radiocarbon. We have already discussed the way in which

C-14 is generated and mixed with the carbon in the atmosphere–ocean system. Foram shells and coccolith tests when formed in surface ocean water contain C-14 atoms as well as ordinary C atoms. Upon formation, they contain about one C-14 atom per 10^{12} C atoms (although very rare, this C-14 is present in enough quantity to be measured). Once formed, the shells constitute an isolated system; their C-14 content decreases with time in the logarithmic manner described previously: half of the C-14 disappears during the first 5700 years, half of the remainder disappears during the next 5700 years, and so on.

The fundamental assumption of radiocarbon dating is that the C-14/C ratio in sea water has always been the same as it is today. If this is so, then a shell sample from a deep sea core yielding a C-14/C ratio half that in present-day sea water carbon must have formed 5700 years ago. If we find the level in the core where the ratio of C-14/C is only 25 percent of that in the present surface ocean, the sample must have formed 11,400 years ago, and so on. At an age of about 40,000 years, less than 1 percent of the original C-14 remains. Beyond this point, the exchange of carbon in the shells with their surroundings produces enough contamination to make the method unreliable. A significant fraction of the C-14 in shells older than this limit comes from secondarily introduced carbon. Although measurements can be made back to 60,000 years, the method is really not very reliable for marine carbonates beyond 40,000 years.

Although the halving method is an easy way to envision what is happening, it is only in rare cases that our sample will have aged some multiple of 5700 years. If we want to make actual age calculations, we need an equation which relates the C-14/C ratio to time, or:

$$\left(\frac{\text{C-14}}{\text{C}}\right)_{\text{today}} = \left(\frac{\text{C-14}}{\text{C}}\right)_{\text{formation}} \times e^{-\lambda t}$$

This equation states that the C-14/C ratio in the shell at any given time t after formation is equal to the original C-14/C ratio times e (the base of the natural logarithmic system) to the power $-\lambda t$, where λ is the decay constant of C-14 (1 part disappears every 8200 years). Solving this equation for t, we obtain:

$$t = 8200 \ln \frac{(\text{C-14/C})_{\text{formation}}}{(\text{C-14/C})_{\text{today}}}$$

The age t of the sample is equal to 8200 years (the average lifetime of a C-14 atom) times the natural logarithm (ln) of the ratio of the C-14/C ratio at the time of formation to the C-14/C ratio found in the sample today. We assume that $(\text{C-14/C})_{\text{formation}}$ is identical to the ratio we find in present surface ocean water (and hence in living shells and coccoliths).

Uranium Series Dating

The second dating method involves the element uranium (U) and two radioactive "daughter" isotopes produced by the decay of uranium, Th-230 and Pa-231. (See Figure 3-8 for the position of these isotopes in the decay chains of U-238 and U-235.) We will concentrate first on Th-230, the product of long-lived U-238. Uranium-238 was produced from hydrogen in the interior of stars that subsequently exploded and dispersed their products within our galaxy. It was incorporated into the solar system when it formed from a galactic dust and gas cloud about 4.5 billion years ago. Uranium-238 is unstable (that is, radioactive) and has been gradually decaying since its stellar birth. Since a uranium atom has a mean life of about 7 billion years, however, not enough time has elapsed for U-238 to disappear entirely and a fair amount of it remains in our solar system. Uranium-238 gradually decays through a whole series of radioactive forms to stable lead (Pb-206). Many billions of years from now all the U-238 will have been converted into Pb-206. Until then, we will have in nature a whole series of uranium decay products, produced as uranium atoms and decaying towards lead (see Figure 3-8, pages 76–77).

The transition from U-238 to Pb-206 requires the emission of 8 alpha particles, each carrying away 2 neutrons and 2 protons. It also requires an equivalent number of electron emissions. Obviously, an atom must go through many stages in the transition from U-238 to Pb-206, and the most interesting step is its pause at Th-230. This isotope has a half-life of 75,000 years (ideal for dating in the range of a few hundred thousand years). Fortunately for dating purposes, Th-230 has a much different chemical cycle in the ocean than uranium does. Uranium is quite soluble, as indicated by its mean residence time in the ocean of about 600 mixing cycles, or 1,000,000 years. Its concentration is thus relatively high with respect to its abundance in natural rock. On the other hand, thorium is extremely reactive in sea water: it is removed in a matter of a few months. When U atoms dissolved in sea water undergo radioactive decay and produce Th-230, the Th-230 atoms are soon extracted from the sea water onto particulate matter destined for the sediments. This separation between parent U-238 and daughter Th-230 is the basis for uranium dating. The age of the sediments can be computed from the disappearance after burial of the Th-230 received from the uranium dissolved in the sea, in much the same way as C-14 ages are computed. However, successful dating by the Th-230 method is possible only under special circumstances: both the rate at which the sediment particles fall onto a given area of the sea floor and the rate at which the Th-230 atoms are incorporated into this falling sediment must remain constant with time. The trick is to know whether these assumptions are valid for any given core. (Fortunately, this can be determined, as we will see later in this chapter.) If the core meets these criteria, we use the Th-230 content to calculate ages; if it does not, we disregard the measurements. Many of the sediments deposited in the open ocean do accumulate at a sufficiently constant rate to make reliable dating by this method possible.

Just as with the radiocarbon method, the amount of Th-230 in a given weight of sediment at any given depth in an acceptable core is equal to the amount of Th-230 in the sediment when it was deposited times $e^{-\lambda t}$. For Th-230, one part in 108,000 disappears each year. If our assumptions are correct, then every unit of sediment in that core began with the same amount of Th-230, and the amount that remains is a measure of the time that has elapsed since the sediment formed.

Now we will return to evaluating the reliability of the Th-230 method on a given core. Figure 3-9(a) is a plot of the fraction of Th-230 remaining as a function of depth in an ideal core that has a constant sedimentation rate of 2 cm/10^3 yrs. If we measured the Th-230 content of this material, we would find that at a depth of 150 centimeters (or 75,000 years in age) the amount of Th-230 had dropped by a factor of 2 from that found at the top of the core. If we then took a sample at 300 centimeters (150,000 years in age), we would find only one-fourth of the original Th-230 content. If we went down to 450 centimeters, the age of the sample would be 225,000 years (three half-lives would have passed) and only one-eighth of the initial Th-230 would remain. For an ideal core, the Th-230 concentration should follow an exponential curve with depth. In order to test whether points fall along such a curve, it is best to transform the coordinates of the graph in such a way that they lie along a straight line. This can be accomplished by plotting the logarithm of the Th-230 concentration (or the logarithm of the ratio of Th-230 at any given depth to the amount at the surface) against depth. If the core meets the criteria of the method then the points should fall along a straight line, as they do in Figure 3-9(b).

Figure 3-10(a) plots an actual set of Th-230 measurements on a core from the Caribbean Sea. This core proves to be close to ideal. Although there is some scatter from a single straight line, most of the points lie within twice the measurement error of this line. The best fit line corresponds to a sedimentation rate for the core of 2.35 cm/10^3 yrs. To show the sensitivity of the sedimentation rate to the choice of the best fit line, solid lines representing sedimentation rates of 3.00 cm/10^3 yrs (25 percent higher) and 1.90 cm/10^3 yrs (25 percent lower) have been included for comparison. Even in the best cores, this method allows the sedimentation rates to be determined only within a range of 10 or 15 percent. Individual age determinations calculated by dividing the depth assigned to a given event by this average rate could be in error by even greater amounts.

The other isotopic product of uranium decay, Pa-231, is being

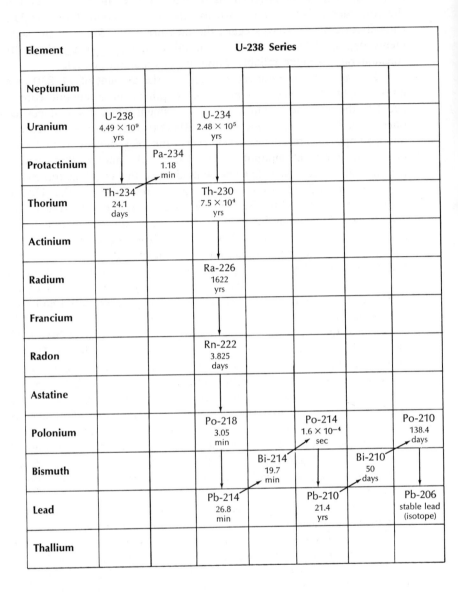

Element	U-238 Series						
Neptunium							
Uranium	U-238 4.49 × 10⁹ yrs		U-234 2.48 × 10⁵ yrs				
Protactinium		Pa-234 1.18 min					
Thorium	Th-234 24.1 days		Th-230 7.5 × 10⁴ yrs				
Actinium							
Radium			Ra-226 1622 yrs				
Francium							
Radon			Rn-222 3.825 days				
Astatine							
Polonium			Po-218 3.05 min		Po-214 1.6 × 10⁻⁴ sec		Po-210 138.4 days
Bismuth				Bi-214 19.7 min		Bi-210 50 days	
Lead			Pb-214 26.8 min		Pb-210 21.4 yrs		Pb-206 stable lead (isotope)
Thallium							

Figure 3-8   The decay chains of the three long-lived alpha particle emitting radioisotopes found on the earth. The half-life of each of the "daughter" products formed is given. The vertical arrows represent alpha particle emissions and the diagonal arrows represent electron emissions. In addition to Th-230 and Pa-231,

Th-232 Series					U-235 Series				
					U-235 7.13×10^8 yrs				
						Pa-231 3.25×10^4 yrs			
Th-232 1.39×10^{10} yrs		Th-228 1.90 yrs			Th-231 25.6 hrs		Th-227 18.6 days		
	Ac-228 6.13 hrs					Ac-227 22.0 yrs			
Ra-228 6.7 yrs		Ra-224 3.64 days					Ra-223 11.1 days		
		Rn-220 54.5 sec					Rn-219 3.92 sec		
		Po-216 .158 sec	65%	Po-212 3.0×10^{-7} sec			Po-215 1.83×10^{-3} sec		
			Bi-212 60.5 min					Bi-211 2.16 min	
		Pb-212 10.6 hrs	35%	Pb-208 stable lead (isotope)			Pb-211 36.1 min		Pb-207 stable lead (isotope)
			Tl-208 3.1 min					Tl-207 4.79 min	

which are used for dating sediments, we will see in Chapter 6 that Ra-226 and Rn-222 in the U-238 series and Ra-228 and Th-228 in the Th-232 series are important indicators of the rates of processes taking place within the sea. Each series ends with a stable isotope of lead.

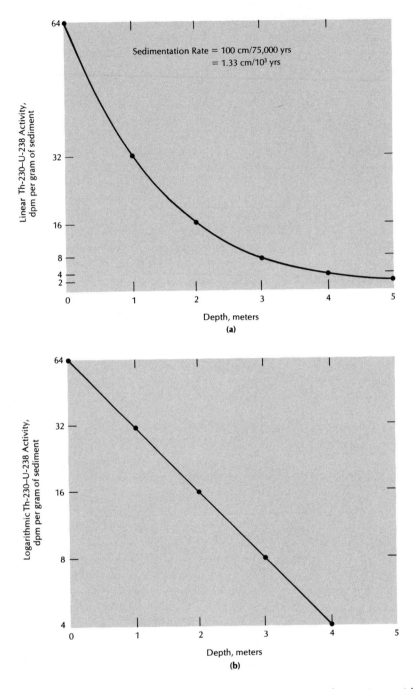

Figure 3-9 The distribution with depth of excess Th-230 in an ideal core with a constant sedimentation rate of 1.33 cm/10³ yrs and a constant rate of Th-230 influx. In such a core, the amount of excess Th-230 shows a drop by a factor of 2 for each interval of 100 centimeters penetrated. (This thickness of sediment is deposited in one half-life of Th-230.) (a) employs a linear scale for the Th-230 content; (b) employs a logarithmic scale. (Note that the points in (b) fall along a straight line.)

produced by the decay of a second isotope of uranium present in sea water, U-235. Like U-238, U-235 has a very long half-life (.7 billion years). Uranium-235 decays to Pb-207 (stable lead), and like U-238, it must go through many stages to make the transition. One stopping point is Pa-231, an isotope with a half-life of 34,000 years. Like Th-230, Pa-231 is very insoluble and is removed from sea water to the sediments soon after it is produced. We can measure the concentration of the isotope Pa-231 at various depths in a deep sea core and plot these data on a logarithmic graph in the same manner we did with Th-230. If the points yield a straight line on such a plot, then the slope of that line indicates the sedimentation rate for the particular core. Figure 3-10(b) plots the Pa-231 measurements for the Caribbean core just discussed. Since Pa-231 is less abundant initially in the core top and disappears more rapidly than Th-230, Pa-231 can be measured with sufficient accuracy only to a depth of 3 meters (compared to a depth of 9 meters for Th-230). Like Th-230, the best fit line for Pa-231 has a slope that dictates a sedimentation rate of 2.35 cm/10^3 yrs! So, in this particular core, both isotopes tell us that the sediment accumulated at roughly a constant rate. Thorium-230 allows the mean rate to be established over an interval of 400,000 years; Pa-231, over an interval of 150,000 years.

Potassium–Argon Dating

To date beyond 400,000 years, we must turn to the potassium–argon method. Potassium-40 is another of the isotopes generated in long ago stars and later incorporated in our solar system. Its half-life is just over a billion years, so although it is also destined for extinction, some K-40 still remains in our earth. Roughly 90 out of 100 K-40 atoms change themselves into Ca-40; the remaining 10 change themselves into Ar-40 (see Figure 3-11). The proportion of K-40 undergoing decay by each of these routes is a constant. When age determinations are made, it is the argon produced by the decay of potassium that is measured rather than the calcium. The reason for this is that when volcanic rock first forms, gases are almost totally excluded from the minerals that crystallize out of the hot liquids. Hence volcanic rocks are initially free of Ar-40. With time, the K-40 in these rocks undergoes radioactive decay and produces argon, which remains trapped in the crystal lattices of the rock. If we measure the amount of Ar-40 and the amount of K-40 in one of these rocks, it is possible to calculate the length of time that has elapsed since the rock crystallized.

To apply this method directly to deep sea sediments requires the chance presence of a layer of volcanic ash. The normal K-bearing minerals found in sediments are detrital in origin and bear the great age of

Figure 3-10   Actual Th-230 (a) and Pa-231 (b) measurements for core V12-122 raised by the Columbia University research vessel *Vema* in the Caribbean Sea. The failure of the points to fall along a single straight line in either graph reflects a combination of experimental uncertainty in the measurements (note error bars) and of nonideality in the core (the sedimentation rate and/or the isotope influx rate did not remain exactly constant with time). The best fit line (a rate of 2.35 cm/10^3 yrs) and lines representing sedimentation rates 25 percent greater (3.00 cm/10^3 yrs) and 25 percent smaller (1.90 cm/10^3 yrs) are shown for both isotopes. Clearly, with the observed scatter, the rate (and hence the age at any given depth) cannot be determined to better than ±10 percent. The agreement between the best fit rates for Th-230 and Pa-231 adds support to the validity of the ages derived in this way.

the rocks from which they were eroded. Unfortunately, volcanic ashes in deep sea sediments are quite rare, so the potassium–argon method does not prove to be very profitable in dating sediments. The few cores dated in this way, however, do yield sedimentation rates in agreement with those obtained by other dating methods.

The real power of the potassium–argon method lies in its use in establishing the chronology of magnetic reversals. Volcanic rocks believed to have formed in the last few million years were collected from all over the earth for this purpose. For each sample, the polarity of the magnetic field at the time that the rock crystallized was established by measuring its magnetic vector. This was possible because the Fe-bearing minerals in basaltic rocks were magnetized in the direction of the North Pole at the time the rocks crystallized and cooled. If a rock crystallized during an interval when the magnetic field was reversed, then today its magnetic vector would point toward the South Pole instead of the North Pole. Roughly half the rocks collected fell in the North Pole category and half fell in the South Pole category. The potassium–argon age of each rock was then carefully measured. This dating established the magnetic chronology for earth: virtually all the rocks with ages of less than .7 million years (700,000 years) were normal in polarity; most of the rocks between .7 and 2.5 million years had reversed polarity, except for two short time intervals during which earth's magnetic field returned to normal (see Figure 3-6).

A few years ago it was found that the same temporal sequence of magnetic polarity observed in volcanic rocks existed in deep sea cores. The orientation of magnetic grains in a core was in accord with the earth's field at the time the sediment was deposited. Once it was established that deep sea sediments record magnetic changes, it was possible to assign ages to deep sea cores based on the volcanic rock measurements already made. We have a means of correlating the events in deep sea cores with the carefully dated sequence of volcanic rocks, provided

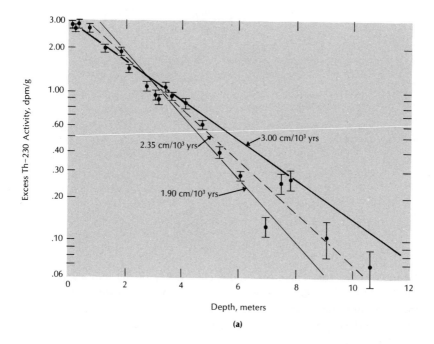

(a)

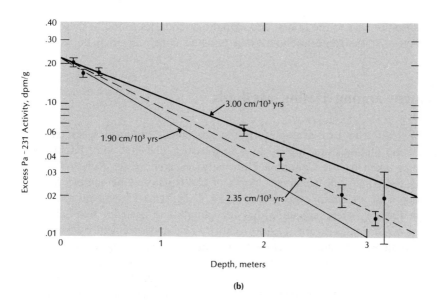

(b)

a core penetrates deep enough to reach at least the youngest magnetic change. Magnetic correlation dating allows us to go much further back in time than other dating methods (to about 4,000,000 years).

Once the chronology of a series of cores has been defined by one or more of the methods just described, it is then possible to assign

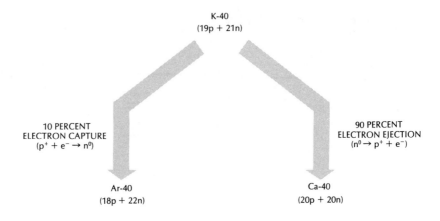

K-40
(19p + 21n)

10 PERCENT
ELECTRON CAPTURE
$(p^+ + e^- \rightarrow n^0)$

90 PERCENT
ELECTRON EJECTION
$(n^0 \rightarrow p^+ + e^-)$

Ar-40
(18p + 22n)

Ca-40
(20p + 20n)

$t_{1/2} = 1.1 \times 10^9$ yrs

Figure 3-11   The scheme for radioactive decay of the long-lived potassium isotope, K-40. One nucleus in ten transforms by capturing an electron; the other nine transform by emitting an electron. The argon isotope, Ar-40, produced by the electron capture process provides a means of dating volcanic rocks.

absolute ages to the faunal and climatic boundaries found in these cores. Ages for the same boundaries in undated cores can then be assigned.

Agreement among Dating Methods

We will cite two examples of the agreement obtained when several dating methods are applied to a single core. The first core, RC11-209, was taken by the Columbia University research vessel *Robert Conrad* in the equatorial ocean just north of the equator. One sample from a depth of 11 centimeters had a C-14/C-12 ratio 25 percent the ratio in surface water, indicating that the sample had aged for two half-lives of C-14 (that is, for 11,400 years). This measurement suggested a sedimentation rate very close to 1.0 cm/10^3 yrs. In this core, the first magnetic boundary from normal (above) to reverse (below) occurred at 700 centimeters. As this reversal has been dated by the potassium–argon method to be 700,000 years in age, the average sedimentation rate was again 1.0 cm/10^3 yrs. By measuring the variation in $CaCO_3$ content with depth in the core, it was possible to establish the climatic record and, particularly, the point at which the second sharp warming before present occurred. In other cores, this boundary has been determined by the Th-230 method to have an age of 125,000 years. The use of this time marker yielded a sedimentation rate for the core of about 1.2 cm/10^3 yrs. So the three methods—direct C-14 dating, the magnetic correlation of an event dated by the potassium–argon method, and the climatic correlation of an event dated by the Th-230 method—all gave roughly the same result for the same core.

The second core, V12-122, was the Caribbean core already used as an example of Th-230 and Pa-231 dating (see Figure 3-10). These two methods yielded concordant rates of 2.35 cm/10^3 yrs. At 9 meters depth in this core, there was a faunal change that could be correlated from core to core. This change has been dated at 400,000 years by interpolating between the core top and the level of the first magnetic reversal (700,000 years) in 5 cores for which a magnetic stratigraphy has been established and in which this particular faunal boundary has been identified. If we adopt this age for a level of 900 centimeters in core V12-122, we obtain a sedimentation rate of 2.25 cm/10^3 yrs. In the same core, C-14 ages of 8950 years at 27 centimeters and 18,300 years at 57 centimeters have been obtained. These ages suggest a slightly higher sedimentation rate of 2.7 cm/10^3 yrs. However, rates based on K-40–Ar-40, Th-230, Pa-231, and C-14 dating methods are remarkably consistent.

Unfortunately the results obtained by these methods are not always this good. Especially for carbonates older than 20,000 years, the C-14 method may be subject to serious errors caused by contamination with recent carbon. Many cores do not yield the ideal Th-230 versus depth curves and hence are not datable by this method. Ash layers may be contaminated with "old" material torn from the throat of a volcano and hence give anomalous ages. The important point to be made here is that cores have been found which *do* give demonstrably reliable results and that these results have provided the basis for the absolute chronology established for the deep sea stratigraphic record.

Measured CaCO₃ Accumulation Rates

Now if we consider the accumulation rates for $CaCO_3$ obtained as a result of dating many deep sea cores throughout the ocean (see Figure 3-12), we find that the average $CaCO_3$ accumulation rate is on the order of .8 g/cm^2 10^3 yrs. The $CaCO_3$ rate in each core is obtained from a knowledge of the bulk rate of accumulation, the percent of $CaCO_3$, and the mean density (on a dry basis). Although three times lower than our predicted value of 2.5 g/cm^2 10^3 yrs, the agreement at least in magnitude between these widely divergent approaches is encouraging.

The difference between the two results may be explained in several ways. First, the rate of vertical mixing between surface and deep water may be only half as great as our two-box model of oceanic mixing suggests. If instead of a 200 cm/yr upwelling rate we used only a

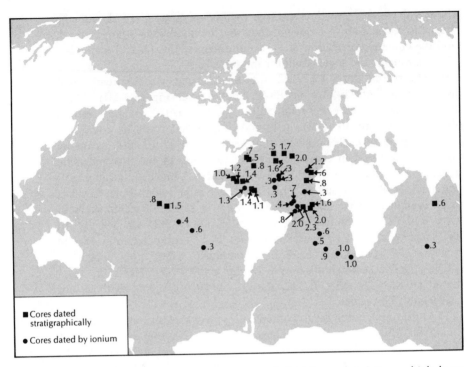

Figure 3-12 Map showing the actual rates of CaCO₃ accumulation which have been obtained by absolute dating of deep sea cores of known CaCO₃ content. The rates are given in units of g/cm² 10³yrs. The two-box model of oceanic mixing predicts an average rate of 2.5 g/cm² 10³ yrs in areas where dissolution has not altered the falling CaCO₃. Asterisks designate those cores where dissolution effects can be demonstrated.

100 cm/yr rate, then the predicted CaCO₃ accumulation rate (1.2 g/cm² 10³ yrs) would be much closer to the observed value. As we will see in Chapter 4, one of the basic weaknesses in the two-box model is that it leads to an excessive estimate of the overturn rate. On the other hand, it may be that in areas where calcite ooze is accumulating, roughly half the CaCO₃ which falls is dissolved. (None of the cores dated, for example, contained any aragonite because they were all collected below the saturation horizon for this metastable mineral.) Also some calcite might be destroyed as it passes through animal viscera. So to the possibility that our calculated overturn rate is too high, we must add the possibility that no place on the sea floor provides a true measure of the rate of CaCO₃ rain into the deep sea. At this point, we cannot state whether either of these explanations is acceptable. Nevertheless the fact that these two independent estimates of the CaCO₃ accumulation rate are no further apart indicates that we really have a fair idea of how rapidly the oceanic mill is grinding.

Calcium Input Compared with Calcium Loss

One other cross-check of the CaCO₃ accumulation rate can be made: we can compare the rate at which $CaCO_3$ is being removed from the ocean with the rate at which calcium is being added to the ocean from rivers. If the oceanic system is roughly at steady state, then the rate at which Ca is being removed in the form of $CaCO_3$ ought to match the rate at which Ca is being added to the ocean. The amount of Ca entering the ocean is given by the product of the mean Ca content of river water ($.10$ moles/m³) times the volume of river water entering the sea ($.10$ meters $\times$ m³/yr A_{ocean}). The amount of Ca leaving the ocean is obtained by multiplying the mean accumulation rate of $CaCO_3$ ($.08$ moles/m² yr) times the fraction of the ocean floor on which $CaCO_3$ accumulates ($.20\ A_{ocean}$). Thus we have

$$\text{entering Ca} = .010\ A_{ocean}\ \text{moles/m}^2\ \text{yr}$$

$$\text{departing Ca} = .016\ A_{ocean}\ \text{moles/m}^2\ \text{yr}$$

The near agreement between these two fluxes supports the basic reliability of the assumption that the system is operating at steady state.

Summary

In this chapter we have seen that the three independent means of establishing the absolute rates of oceanic processes are self-consistent. The rate of river influx was estimated from river gauge records. The rate of vertical mixing was derived from the difference in the C-14/C ratio in average surface and average deep Pacific water. The rate of $CaCO_3$ accumulation was obtained by absolute age determinations made by using several radioisotopes on deep sea sediments of measured $CaCO_3$ content. Two cross-checks were conducted: a comparison of the observed $CaCO_3$ accumulation rate with that predicted from the two-box model of oceanic mixing, and a comparison of the rate of Ca input from rivers with the rate of loss to $CaCO_3$ in the sediments. Even though the resultant river flow data are for the last few decades, the vertical mixing rate data are for the last few thousand years, and the sediment accumulation rate data are for the last few tens of thousands of years, the results are surprisingly self-consistent.

The fact that deep sea sediments have accumulated at remarkably similar rates over the last few million years (indicated by the potassium–argon calibrated magnetic reversal chronology), as compared with the past few hundred thousand years (indicated by the Th-230 and Pa-231 dating methods), and as compared with the past few tens of thousands of years (indicated by C-14 dating), suggests that the present mode of operation is not atypical. The mill has been grinding at nearly the same rate for at least 1,000,000 years!

SUPPLEMENTARY READINGS

■ Publications dealing with the distribution of radiocarbon in the sea:

Brodie, J. W., and Burling, R. W. "Age Determinations of Southern Ocean Waters." *Nature* 181(1958):107–108. Discusses the distribution of C-14 in waters of the South Pacific and Antarctic Ocean.

Broecker, Wallace S., Gerard, R., Ewing, M., and Heezen, B. C. "Natural Radiocarbon in the Atlantic Ocean." *J. Geophys. Res.* 65(1960):2903–31. Discusses the distribution of C-14 in waters of the Atlantic Ocean.

Bien, G. S., Rakestraw, N. W., and Suess, H. E. "Radiocarbon Dating of Deep Water of the Pacific and Indian Oceans." *Bull. Inst. Oceanogr. Monaco* 61 (1963):1278. Discusses the distribution of C-14 in waters of the Pacific and Indian Oceans.

Broecker, Wallace S., and Li, Yuan-Hui. "Interchange of Water between the Major Oceans." *J. Geophys. Res.* 75(1970): 3545–52. A more complex box model of the oceans than that presented in this chapter.

■ Publications dealing with radioisotope dating methods for deep sea cores and the means of stratigraphic correlation for these sediments:

Libby, W. F. *Radiocarbon Dating,* Second Edition. Chicago: University of Chicago Press, 1955. A book covering the principles of radiocarbon dating.

Dalrymple, G. Brent, and Lanphere, Marvin A. *Potassium–Argon Dating: Principles, Techniques, and Applications to Geochronology.* San Francisco: W. H. Freeman & Company, 1969. A book concerned with the principles of potassium–argon dating and the magnetic stratigraphy based on this dating.

Broecker, Wallace S., and van Donk, Jan. "Insolation Changes, Ice Volumes, and the O-18 Record in Deep-Sea Cores." *Reviews of Geophys. and Space Phys.* 8(1970): 169–98. A discussion of the absolute dating of climatic events that emphasizes the results obtained by the Th-230 and the Pa-231 dating methods.

Berggren, W. A. "A Cenozoic Time Scale—Some Implications for Regional Geology and Paleobiogeography." *Lethaia* 5 (Oslo, 1972): 195–215. A discussion of the faunal and floral criteria for establishing useful horizons for global stratigraphic correlation.

■ Books dealing with the global water balance:

Leopold, Luna B. *Water: A Primer.* San Francisco: W. H. Freeman & Company, 1974. A general treatment of the subject of water resources.

Donn, William. *Meteorology*, Third Edition. New York: McGraw-Hill, 1965. A basic meteorology textbook with an excellent coverage of the global water balance.

3-1 Two large water samples from the same location and depth in the deep Pacific are available for C-14 measurement. If one sample was collected in 1890 and the other in 1970, by how much should their C-14/C ratios differ?

3-2 Planet *X* has an ocean with a stratification similar to ours but with a mean depth of 8000 meters. Measurements reveal that its surface water has a C-14/C ratio twice that in its deep water. What is the average residence time of water beneath the thermocline of this ocean?

3-3 Sediment taken at a depth of 50 centimeters in a deep sea core contains foram shells with a C-14/C ratio 12½ percent that for sediment from a depth of 10 centimeters. What is the apparent sedimentation rate?

3-4 The following results were obtained using uranium series measurements at various depths in a deep sea core. The units are disintegrations per minute per gram (dpm/g) of sediment.

Depth, centimeters	U-238	U-234	Th-230
2	1.3	1.4	65.5
22	1.5	1.6	33.6
42	1.4	1.3	17.4
62	1.2	1.3	9.6
82	1.5	1.5	5.7
102	1.2	1.3	3.5
122	1.4	1.5	2.4

What is the apparent sedimentation rate? If the Pa-231 activity is 6.9 at 2 centimeters, what should it be at 42 centimeters?

3-5 The first change from a normal to a reversed magnetic field is found at 14 meters depth in a core. If the sedimentation rate has been roughly constant, over what depth range could the C-14 method be used to reliably date this core? Over what depth range could the Th-230 method be used to obtain an average sedimentation rate?

4

THE GREAT MANGANESE NODULE MYSTERY

One of the major puzzles of marine chemistry is the presence of objects called manganese nodules scattered in great numbers over large parts of the sea floor. These nodules consist of about 2 parts manganese dioxide (MnO_2) and 1 part iron oxide (Fe_2O_3) and contain 1–3 percent by weight of the oxides of nickel (Ni), copper (Cu), and cobalt (Co). As nickel and copper are in short supply, the great abundance of these sea floor nodules has attracted the attention of the major mining companies, and schemes to "vacuum" them from the deep sea floor are being given serious study.

One mystery with regard to these objects is how they can possibly form when the constituent elements (Fe, Mn, Ni, Co, Cu, · · ·) are in such minute abundance in sea water. In most cases, the concentrations of these elements are so low that techniques to measure them reliably are still under development. Those measurements which have been made were plagued by contamination. Was the metal measured present in the sea or was it added from the sampling device, the storage container, the laboratory glassware and reagents, or another source? So here we have elements with such low abundances in sea water that they defy accurate measurement forming the fourth most abundant material found on the sea floor (after alumino-silicate debris, calcite, and opal). Although this mystery is far from solved, a number of the major clues that have been uncovered will be discussed in this chapter.

As so little is known about the differences in the concentrations of Mn, Fe, Ni, Cu, and Co from one place in the ocean to another, we will not be able to take the same approach here that we did for phosphorus earlier. The evidence from sea water itself is much less direct. We will have to rely on information gleaned from elements other than those that comprise manganese nodules.

A Lesson from Cesium and Barium

First let us consider the cycles of two trace metals in the sea for which we do have some reliable concentration data, cesium (Cs) and barium (Ba). Although both of these elements have relatively short residence times in the sea (~40,000 years) and hence must be quite reactive, they show very different concentration patterns. Cesium shows *no* measurable concentration differences between surface and deep water (see Figure 4-1(a)), while barium shows a threefold higher concentration in deep Pacific than in surface Pacific water (Figure 4-1(b)).

Barium must follow the same kind of cycle as C, P, N, Si, Ca, $\cdots$. (Like carbon and calcium, barium's surface value drops until the nutrient elements have been consumed and then remains constant.) Since the product of Ba^{++} and SO_4^{--} ion concentrations in the sea is close to the solubility limit of the mineral barite ($BaSO_4$), it has been suggested (but not directly demonstrated) that organisms manufacture this salt. If this is so, $BaSO_4$ must be formed in surface water and largely (but not completely) destroyed in the deep sea.

Cesium has no such insoluble salt. The somewhat disturbing point arises that Cs can escape from the sea in a time as short as 25 mixing cycles without leaving any concentration differences to bear witness to its mode of departure! This means that the low abundances of the metals Fe, Mn, Ni, Co, Cu, $\cdots$ in the sea need not be reflected by the large surface to deep water differences indicative of intense biological uptake.

Figure 4-1   Distribution of dissolved cesium (a) and dissolved barium (b) with depth in the Pacific Ocean (28° N, 122° W). In order to eliminate the small variations related to salinity changes, the cesium data have been recalculated to give the concentrations which would have been found if all the samples were brought to 35.00 per mil salt content. (Data for cesium obtained by T. Folsom, Scripps Institution of Oceanography, and data for barium obtained by K. Wolgemuth, Lamont-Doherty Geological Observatory.)

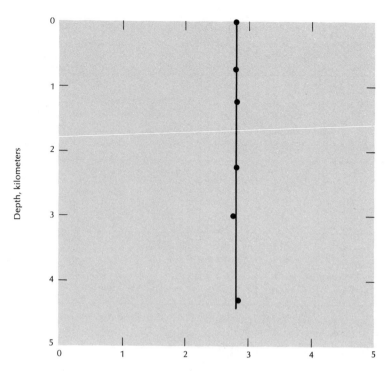

Cesium, 10^{-7} g/liter
(normalized to 35.00 per mil salinity)

(a)

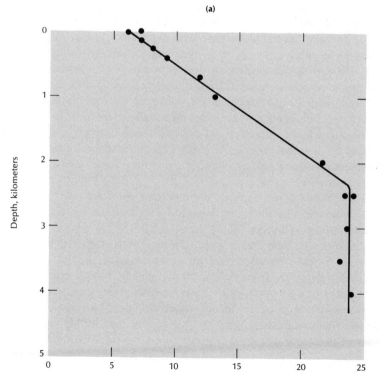

Barium, 10^{-6} g/liter

(b)

Table 4-1   Alkali and alkaline-earth element abundances in mean river and sea water, and the residence times calculated from these abundances.

	Element	Atomic Weight, g/mole	Ionic Radii, 10^{-8} cm	Average Concentration, mg/liter		Residence Time, years
				Sea	River	
Alkali	Na	23.0	.95	10,000	2*	200,000,000
	K	39.1	1.33	380	2	8,000,000
	Rb	85.5	1.48	.12	.03	120,000
	Cs	132.9	1.65	.0003	.0003	40,000
Alkaline Earth	Mg	24.3	.65	1350	1.3	40,000,000
	Ca	40.1	.99	400	8	2,000,000
	Sr	87.6	1.13	8	.03	12,000,000
	Ba	137.4	1.35	.025	.050	20,000

* A correction has been made for the contribution of Na from sea salt blown off the ocean and deposited by rainfall on the continents.

Examination of the abundances of the other three alkali metals—sodium (Na), potassium (K), and rubidium (Rb)—reveals that, like cesium, they show no concentration differences in the sea. As we can see from Table 4-1, Na, K, Rb, and Cs also show a systematic (and large) *increase* in residence time with *decreasing* atomic radius. (Their reactivity must rise rapidly with ionic size.) Although no one has been able to show the exact mechanisms of removal, the absence of a concentration pattern generated by the biological cycle and the increase in reactivity with size suggest that absorption onto silicate minerals may be occurring. This could take place through interaction with growing authigenic silicate phases, land-derived silicate debris, or weathering submarine volcanic rocks. Any of these interactions could yield the observed reactivity pattern without generating large concentration differences within the sea.

The latter point requires some careful consideration. Two models of Cs removal are shown in Figure 4-2: (a) assumes Cs removal in surface waters (that is, along continental shelves and in estuaries); (b) assumes Cs removal at great depths (onto deep sea sediments or volcanic rocks). Provided that in either case the cesium entering the oceans is primarily from rivers, then if model (a) is correct, surface water and deep water will have identical Cs concentrations, but if model (b) is correct, surface water will be 5 percent richer in Cs content than deep water. The absence of any measurable vertical variations at the 1 percent level suggests that Cs is removed primarily from

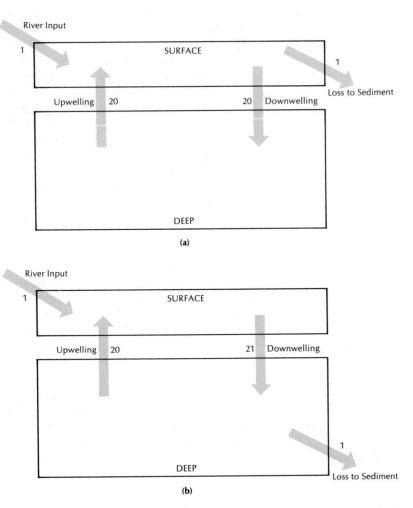

Figure 4-2 Two models for the cycle of cesium within the sea. In (a), Cs is added to and removed from the surface reservoir. In such a system, there would be no difference between the Cs content of surface and deep water. In (b), Cs is again added to the surface reservoir but it is removed from the deep reservoir. The Cs content of surface water must therefore be 5 percent higher than the Cs content of deep water. All fluxes are given as multiples of the rate at which Cs is entering the oceans from rivers.

the surface ocean rather than from the deep sea. The residence times of the other three alkali metals are so great that no measurable differences in concentrations at even the .1 percent level can be expected regardless of the reservoir from which removal occurs.

Table 4-1 also lists the residence times for three of barium's sister alkaline earths: magnesium (Mg), calcium (Ca), and strontium (Sr).

In this case, the simple relationship between residence time and size does not hold. Although larger than Ca, Sr has a residence time five times higher. This complexity is related to the fact that, like Ca, the element Sr appears to be removed from the sea in the form of calcite. Measurements of the Sr/Ca ratio in the coccolith tests and foram shells found in deep sea sediment reveal that the organisms producing this calcite discriminate against Sr. The Sr/Ca ratio in calcite averages about one-sixth that in sea water, but, interestingly enough, the Sr/Ca ratio in calcite is nearly the same as it is in rivers. Thus we can see why Sr atoms remain in the sea longer than Ca atoms. Since organisms discriminate against Sr by a factor of about 6, the escape probability of Sr is six times smaller than it is for Ca and the residence time of Sr is six times greater! The fact that the Sr/Ca ratio is nearly the same in calcite as it is in rivers is expected if calcite is the major removal agent for both these elements.

Although sufficiently accurate measurements of the Sr concentration within the sea have not been made to demonstrate differences between surface and deep water, we can show from our knowledge of the distribution of Ca and of the Sr/Ca ratios given above that surface Pacific water should contain about .2 percent less Sr per gram of salt than deep Pacific water. (Because of discrimination by organisms, the difference between the Sr content of surface and deep water should be six times smaller than the difference of 1.2 percent found for Ca.)

So we see that the removal of three of the four alkaline-earth elements is heavily dependent on the biological processes taking place in the sea. The removal of the four alkali metals apparently is not. The lesson to be learned here in connection with predictions about the metals Fe, Mn, Ni, $\cdots$ is that inorganic processes can be effective in removing elements from the sea without generating large chemical inhomogeneities within the sea.

The Rapid Removal of Thorium

Our next lesson comes from the element thorium. This element is of particular interest because it has four radioactive isotopes: Th-228, Th-230, Th-232, and Th-234 (see Figure 3-8). In Chapter 3 we saw that Th-230 is used in marine chemistry to determine sedimentation rates. Two of the other isotopes, Th-228 and Th-234, reveal some exciting information about the rapid removal of highly insoluble substances. Both Th-228 and Th-234 have short half-lives (1.8 years for Th-228, and 25 days for Th-234). Both are produced within the sea by the radioactive decay of their much more soluble parent isotopes

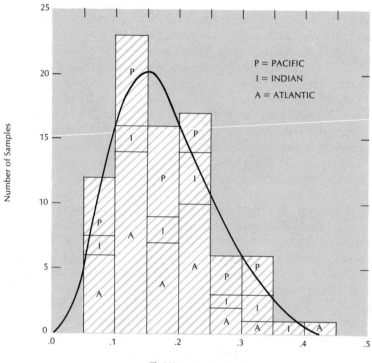

P = PACIFIC

I = INDIAN

A = ATLANTIC

Th-228 Activity/Ra-228 Activity

Figure 4-3 Histogram showing measurements of the activity ratio of Th-228 and its parent Ra-228 in surface ocean water. If thorium were not being rapidly removed by falling particles, the ratio would be expected to be unity. A value of .15 for this ratio corresponds to a chemical removal time of six months for the element thorium. (Data obtained by A. Kaufman, Lamont-Doherty Geological Observatory.)

(Ra-228 in the case of Th-228, and U-238 in the case of Th-234). By finding the ratio of the amount of daughter isotope to the amount of parent isotope at various places in the sea, it is possible to determine the ratio of the removal probability via particles to the removal probability via radioisotopic decay. Since the radiodecay rate is known, the particle removal rate can be calculated from the daughter/parent ratio.

As summarized in Figure 4-3, the ratio of Th-228 to Ra-228 found in surface water to that predicted if particulate removal were not operative averages .15. This very low ratio suggests that particulate removal is *more* important than radiodecay. (If they were equally important, the ratio would average .50.) Since the mean life of Th-228 with respect to radioisotopic decay is 2.6 years, the observed .15 value requires that the mean life of thorium with respect to particulate removal be only .4 years! A Th-228 atom has a seven times greater probability of leaving the surface ocean on a particle than of being transformed by radiodecay within the surface sea.

Although a less precise indicator, a 15 percent deficiency of Th-234 relative to its parent U-238 in surface water also suggests a five-month removal time for Th atoms produced within the surface sea. Since the techniques used to extract thorium from sea water recover the Th present in particulate as well as in dissolved form, this five-month removal time reflects not only the uptake onto the particles but also the removal of particles from the surface water mass.

What process might operate so rapidly in sea water? To see why organisms are the likely villains, let us reconsider the cycle of phosphate in the surface ocean. If the mean thickness of the surface ocean reservoir is taken to be 100 meters, then the 2 m/yr upwelling rate (the v_{mix} calculated in Chapter 3) suggests that the water in this layer is replaced once every 50 years (100 m/2 m/yr). Now as the phosphate content observed in surface water is about 30 times lower than the P content of upwelling deep water, the probability of phosphorus removal by organisms must be 30 times greater than the probability of phosphate removal with downwelling water. The residence time of a P atom at the surface is only one-thirtieth that of a water molecule! Thus the mean residence time of P in surface water must be only 1.4 years. As pointed out earlier, phosphate is recycled about 6 times within the surface ocean before a particle succeeds in carrying it to the deep sea. Thus the time between successive photosynthetic fixations is only about three months. In their quest for limiting building blocks, plants appear to "process" the entire volume of surface ocean water several times each year. It is therefore quite likely that the highly insoluble Th atoms encountered in this search are inadvertently removed by the plants!

Once removed from sea water by a plant, very insoluble elements like Th are unlikely to be released during the consumption of these plants by animal or bacterial life in the sea. Instead they are either passed along the food chain or incorporated into insoluble fecal matter. Thus thorium does not participate in the manifold recycling within the ocean that phosphate enjoys. Instead, Th doggedly remains in the solid phase (animal tissue, fecal pellets, and so on) and is carried in this form to the bottom where it becomes part of the sediment.

Our second lesson, then, is that not all elements taken up by organisms remain in the ocean for a large number of mixing cycles. Those very insoluble substances which are inadvertantly incorporated by organisms in their search for the limiting nutrients (or perhaps absorbed onto their dead remains or excretion products) may well be permanently removed within a matter of months after they enter the sea! The range of residence times can be extended from 2×10^8 years for Na to less than 1 year for Th! It may be that the metals Mn, Fe, Ni, Cu, Co, $\cdot \cdot \cdot$ act more like Th than either Cs or Ba in the sea.

Sources of the Constituents in Manganese Nodules

For our next clue, let us turn our attention to the relationships between the concentrations of the elements making up the mysterious deep sea nodules in continental rocks and in deep sea sediments. We will see from this comparison that these elements show an enrichment not just in nodules but in all marine deposits. Although more spectacular in nodules, the overall enhancement proves to be very significant.

Of the continental rocks found on the surface of the earth, the granitic and basaltic varieties are the most abundant and, for abundance calculations, it is safe to assume that they are the dominant source materials for sediments. A mix of 50 percent granite and 50 percent basalt yields the concentrations of most elements found in sediments. The average abundance of an element in basaltic rock and in granitic rock should then be the value found in average sediment. Now the most abundant sediment is shale; it constitutes about 80 percent of all sediment and is formed from the bulk of the detrital material deposited near continental margins. So the first step is to compare the composition of shale with the basalt-granite average (see Table 4-2). We find that,

Table 4-2   Comparison of the chemical composition of average igneous rock (equal parts granite and basalt) with the chemical compositions of average sediment (shale) and average deep sea sediment (red clay). (Concentrations are in parts per million by weight.) The large excesses of Mn, Ni, Co, Cu, and Fe in red clays (over igneous rock and shale) pose a great mystery.

Element	Granite	Basalt	Average Igneous Rock	Shale	Red Clay
Na	26,000	18,000	22,000	10,000	15,000
K	8,000	42,000	25,000	27,000	25,000
Rb	30	170	100	140	110
Cs	1	4	3	5	6
Mg	46,000	1,600	24,000	15,000	21,000
Ca	76,000	5,000	40,000	22,000	29,000
Sr	470	100	290	300	180
Ba	330	840	590	580	2,300
Fe	86,000	14,000	50,000	47,000	65,000
Mn	1,500	400	950	850	6,700
Ni	130	5	70	70	225
Co	50	1	25	19	74
Cu	90	10	50	45	250
Cr	170	4	90	90	90
Th	4	18	11	12	12

except for the elements Na, Mg, and Ca, shale has the composition obtained by simply mixing together one part basalt and one part granite. This is also true to some extent for deep sea sediments (excluding their opal and calcite components) but, as shown in Table 4-2, deep sea sediments show abundances for a number of elements that are much larger than either the granite-basalt average or the shale content.

Before considering the significance of these excesses, let us examine the reason for the deficiencies of Na, Mg, and Ca in shale and in deep sea sediments. Sodium is in short supply because much of the Na released by weathering remains dissolved in the sea. If we were to take all the sodium out of sea water and mix it with every sediment around, the sediments would be brought close to their expected Na content.

Calcium and magnesium are deficient in shales because they reside in limestone (calcium carbonate) and dolomite (calcium-magnesium carbonate) as well as in shale. We have already seen that $CaCO_3$ is an important primary constituent of marine sediment; $MgCa(CO_3)_2$ is not. The problem of when and where the large amounts of dolomite found in ancient sediments formed constitutes a major controversy in geology. Dolomite is forming today only in a few obscure places such as tidal flats. These occurrences cannot explain the enormous amounts of dolomite found in ancient sediments. Some secondary process must take place after the sediment is buried that causes a transformation to dolomite. Presumably, at some time after the sediments form, half of the calcium in the $CaCO_3$ is replaced by magnesium to make dolomite. Unfortunately, we do not yet know the details of this process. But we do know if we were to add all the dolomite and limestone that has formed to the existing shale, we could account for the missing magnesium and calcium.

When we compare deep sea clays with shale, the concentrations agree fairly well except for the elements found in manganese nodules and the element barium. Note that this enrichment in deep sea clays relative to shale does not apply to all trace metals. Chromium (Cr) and thorium (Th) are examples of metals that do not show any anomaly in deep sea clays.

The Mn content of a mixture of equal amounts of granite and basalt would be about 850 parts per million (ppm). This is the average value found for shales. Yet the Mn content of deep sea sediments, although highly variable, averages about 6000 ppm (a factor of 7 higher than expected). For Ni, an average factor of 3 higher concentration in deep sea sediment than in shale is found; for Co, a factor of 4 higher; for Cu, a factor of 5 higher. Iron shows about 50 percent enrichment in deep sea sediments over what would be expected from the granite-basalt average or from shale.

One common characteristic of the elements showing an enrichment in the sediments of the deep sea is that they have a +2 valence.

Elements such as chromium and thorium that do not show this enrichment have a higher valence than $+2$.

Three major hypotheses have been proposed to explain why deep sea sediments are enriched in Mn, Ni, Co, Cu, and Fe. One suggests that these particular elements are derived from continental weathering (the most obvious source). When rock is broken down during this weathering process, the Mn and Fe tend to associate with very fine, lightweight particles that remain suspended for a long time when they reach the ocean; these elements are therefore spread quite uniformly over much of the ocean floor. The heavier particles, which are relatively free of Mn and Fe, do not travel very far from the edges of the continents and are found primarily in near-shore sediments (shale). Since the Mn concentration in deep sea sediments is about seven times greater than the Mn concentration in the granite-basalt source, deep sea sediments cannot constitute more than 13 percent of all sediment. For example, if deep sea sediments constituted 15 percent of the total sediment, *all* of the manganese would be required to enrich them sevenfold, and no Mn would occur in the remaining 85 percent of the sediment.

How much of the material being removed from the continents travels well away from the continental margins? In major rivers, numerous measurements have been made of the amount of small-particle material (the *suspended load*) and of the amount of large-particle material dancing along the river bottom (the *bed load*). Averages of the detrital fluxes produced by the major rivers in the world yield a fairly good estimate of the present rate at which the continents are being denuded. The estimates range from about 4 to 10 cm/10^3 yrs. We will assume here that about 6 centimeters of continental material is being eroded away every 1000 years. This material comes mainly from mountainous regions. For example, a study of the Amazon Basin shows that virtually all of the erosion of that area takes place in the Andes Mountains and hardly any occurs in the flat, rain forest area. This makes sense. The most wear and tear to the rocks occurs where there is the steepest topographic gradient. Because the erosion rate is so low in the flat areas of the continents, to maintain an overall average of 6 cm/10^3 yrs rate requires that the mountainous regions ($\sim$25 percent of the total) be eroded at an average rate of perhaps 25 cm/10^3 yrs.

If the 6 cm/10^3 yrs of continental debris were uniformly spread over the entire floor of the ocean, what sort of sediment thickness would accumulate every 1000 years? To answer the question we must keep two things in mind. First, we are taking a 6-centimeter layer from 30 percent of the earth's surface and spreading it over 70 percent of the earth's surface. That cuts down the thickness of the sediment. Second, the relative density of dry deep sea sediment is only about .8 g/cm^3 compared with the typical continental rock density of 2.4 g/cm^3. This means that the sediment is fluffed up (it is a mixture of rock fragments and water) and thus is three times thicker in the ocean that it was as rock on the continent.

These two corrections almost balance one another. If the detritus coming out of the rivers were spread uniformly over the ocean floor, we ought to receive $6 \times 3/7 \times 2.4/.8$, or $8 \, cm/10^3$ yrs of accumulation. When we confine ourselves to the deep ocean only (ignoring the continental shelves, shallow ocean basins, and the like), we find an average world sedimentation rate of about $.5 \, cm/10^3$ yrs for alumino-silicate detritus. Only about 6 percent of the debris disgorged by rivers reaches the deep sea! The remaining 94 percent must be deposited in small ocean basins, on continental margins, and elsewhere.

Thus if only, let us say, 40 percent of the Mn derived from weathering were to associate with the 6 percent of the material arriving in the deep sea (leaving 60 percent of the Mn for the 94 percent of the material not reaching the deep sea), we could explain the observed Mn enrichment. Since shales show no depletion in Mn with respect to the granite-basalt average, however, this model strains the evidence. Also, there are those who claim that because of deforestation, agriculture, and so forth, the current denudation rates are far higher than they were before the appearance of civilized man. If this is true, then deep sea sediments must constitute considerably more than 6 percent of the total sediment. This conclusion gains support from the failure to locate an adequate amount of near-shore sediment to account for the high rate of accumulation demanded by the denudation rate of $6 \, cm/10^3$ yrs, making manganese from the continents an even less likely source of the Mn enrichment in deep sea sediments.

The second argument which attempts to account for the enrichment anomalies is based on the phenomenon referred to as *secondary enrichment*. In their search for ores, economic geologists often find that ground water has leached precious metals from the underlying rock and redeposited them in the soil horizon, thus producing a rich deposit. The same sort of enrichment could be occurring in the deep sea. As a given layer of sediment gets buried deeper and deeper, the oxygen gas that kept it aerobic (oxidized) is eventually used up by bacteria chewing away at the little bits of organic material left in the sediment. Once this oxygen is depleted, the sediment becomes *anaerobic*; the aerobic bacteria are replaced by bacteria that use sulfate ion as an oxidizing agent. Under these conditions, Mn will change its oxidation state from $+4$ to $+2$. In its $+4$ valence state, Mn is quite insoluble and tends to remain fixed in the sediments. However, Mn in the $+2$ valence state is considerably more soluble and can move about in the pore fluids.

This hypothesis suggests that once sediment is buried deep enough, reducing conditions set in and release the bound manganese. The released Mn then migrates back toward the top layer of sediment where it encounters oxygen and is again precipitated. Thus the upper sediments would contain the manganese associated with the entire sedimentary column. The main problem with this idea is that no evidence of the consequent Mn-*depleted* sediment deeper down has been found. Deep sea cores ±10 meters in length show the same concentration of bound manganese at the bottom as they do at the top. So if Mn-poor sediment exists, it must lie tens of meters below the surface. However, the diffusion path in this case becomes so long that it is doubtful the enrichment process could operate nearly fast enough to bring about the upward transport of manganese.

The third hypothesis suggests that Mn, Ni, Co, Cu, and Fe are being added to the ocean (along with new crustal material) in the regions of subsea volcanism. It argues that the hot basalts coming up from the earth's interior contact water seeping down from the ocean and saturate this water with many of the constituent elements found in the basalt. (It is true that the elements we see enriched in deep sea sediments are indeed the elements that would dissolve into hot salt water.) This metal-laden water then spills out onto the floor of the ocean, bringing Mn, Ni, Co, Cu, and Fe with it.

Now the main zone of volcanism is associated with the mid-ocean rifts caused by sea floor spreading. If these particular elements are added to the ocean as the result of volcanism, they should enter the sea mainly along the crests of the mid-ocean ridges. Sediment samples from these ridge crests do show extremely high concentrations of manganese and iron oxides. Traverses across these rises reveal that the very high Mn-Fe oxide sediments are confined to a narrow belt at the ridge crests. Traveling away from the crests, there is a rapid depletion in the Mn-Fe oxide content of the sediment.

As shown in Table 4-3, sediments on the ridge crests contain about 40 percent Mn-Fe oxide and 60 percent $CaCO_3$. (Only 1 percent of ridge-crest sediments are alumino-silicate debris.) Fifty kilometers from the crests, the $CaCO_3$ content rises to 90 percent and the Mn-Fe oxide content falls to only 7 percent. Radioisotopic dating shows that the accumulation rates of $CaCO_3$ do not differ greatly; thus the change in

Table 4-3   Chemical composition and accumulation rate data for four cores taken along a traverse down the western flank of the East Pacific Rise.

Core Number	Latitude, ° S	Longitude, ° W	Depth, kilometers	Distance from Ridge Crest, kilometers	Sedimentation Rate, cm/10³ yrs	CaCO₃, percent	Fe + Mn Oxides, percent
V19–54	17°02′	113°54′	2.83	50	1.50	65	~30
V19–61	16°57′	116°18′	3.41	300	.70	90	~ 5
V19–64	16°56′	121°12′	3.54	820	.45	90	~ 3
V19–66	16°24′	127°38′	4.10	1500	.15	10	~25

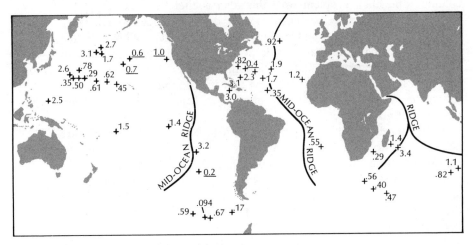

Figure 4-4   Map showing the manganese accumulation rates in mg/cm² 10³ yrs at various points on the sea floor. The rates were obtained by absolute dating of various horizons in the sediment and then determining the average Mn content of the sediment over the intervals dated. Underscored numbers are rates for nodules; other values are for sediments. (Data obtained by M. Bender, Lamont-Doherty Geological Observatory.)

composition reflects a rapid decrease in the rate of Mn-Fe oxide sedimentation away from the ridge crests. Presumably, every part of the sea floor began its history at a ridge crest by receiving 10 or so meters of sediment rich in manganese and iron oxides. As the newly formed crust moved away from the crest, its Mn-Fe supply was rapidly depleted and $CaCO_3$ dominated. By drilling through sediments on the ridge flank, this sequence has been frequently observed. Just above the basaltic crust, the nearly pure calcite ooze gives way to 10 or so meters of high Mn-Fe oxide sediments (see Figures 2-8 and 2-9).

But there is a difficulty with this third hypothesis, too. If mid-ocean ridges are the major source of the manganese in all deep sea sediments, one would suspect that the rate at which Mn is accumulating in sediment would be inversely related to the distance of the sediment from the ridge crests. With the exception of sediments that are right on top of the ridges, this is not the case, however. Elsewhere, the Mn accumulation rate in red clays is fairly uniform over the ocean basins (see Figure 4-4). Of course, it is possible that only a small portion of the manganese rising under the ridge crest is locally precipitated. A much larger amount may escape and be spread widely through the deep sea. But, as with the first two hypotheses, we must stretch the facts if we are to accept the third.

So we have three major hypotheses that attempt to account for the deep sea enrichments of Mn, Ni, Co, Cu, Fe, · · · :

1. separation during continental weathering
2. upward migration within the sediment
3. release by ocean-ridge volcanism

Thus far, the available evidence is not adequate to determine which of these is most likely the dominant source. At this point, intuition gives the volcanism theory the highest rating, but intuition is often misleading.

Growth Rate of Manganese Nodules

Now let us consider the manganese nodules themselves for evidence. Their growth rate provides us with a critical clue. In areas where Mn nodules are found, the accumulation rate of the surrounding sediment averages about .3 cm/10^3 yrs, or 3 *meters* per million years (3 m/10^6 yrs), but the growth rate of the nodules is only one-thousandth as fast, or about 3 *millimeters* per million years (3 mm/10^6 yrs). So a nodule that has a radius of 3 centimeters has taken about 10 million years to grow!

This has been a hard fact for many scientists to swallow, and the validity of the dating methods used to establish this growth rate has been questioned. The original dating was made by using the same Th-230 and Pa-231 methods that were used to determine the accumulation rates of sediments. Rates of 2–4 mm/10^6 yrs were obtained (see Figure 4-5). But the opposition maintained that these dates were not valid because the dating was done entirely on the outer millimeter of these nodules. (Beyond this depth, the Th-230 and Pa-231 excesses completely decay away!) It was suggested that the observed distributions of Th-230 and Pa-231 were generated by adsorption and by other processes which might operate well in the outer millimeter. The fact that the two isotopes gave the same growth rate had to be attributed to chance.

A second method to determine the nodule growth rate was soon found, based on another uranium series isotope, U-234, with a half-life of 254,000 years. Uranium-234 gave the same rate the Th-230 and Pa-231 methods had, with the advantage that a 4-millimeter depth, rather than a depth of only 1 millimeter, was used (see Figure 4-6). Well, 4 millimeters was still small compared to the whole nodule. The critics were not satisfied.

A strong verification of the slow growth rate came when volcanic fragments found at the center of certain Mn nodules were dated by the potassium–argon method. These volcanic fragments, blown into the sea from volcanoes located on oceanic islands, landed on the sea bottom and became nuclei around which the Mn nodules formed. As the time between the volcanic eruption which produced the fragment and the onset of nodule growth is probably quite small, the dates for these

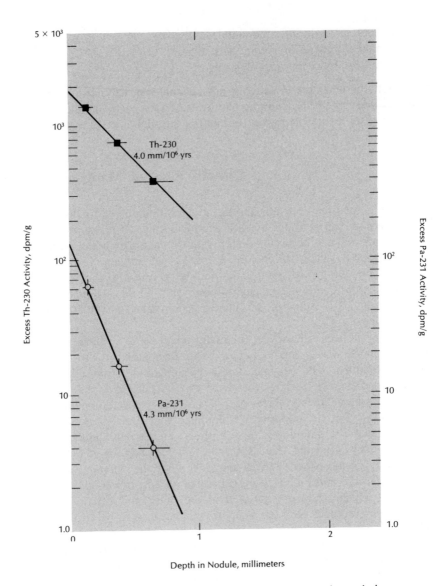

Figure 4-5 Plots versus depth of the natural logarithm of the excess Th-230 activity and excess Pa-231 activity in a nodule dredged from the floor of the North Pacific (36° N, 160° W; 5400-meter depth). The slope of each straight line connecting the points yields the growth rate. In this nodule, both isotopic methods yield a rate close to 4 mm/10⁶ yrs. Note that measurable excesses are only found in the outer millimeter of the nodule. The samples were taken by successively scraping material from a designated area and weighing the scrapings. (Data obtained by T.-L. Ku, Lamont-Doherty Geological Observatory.)

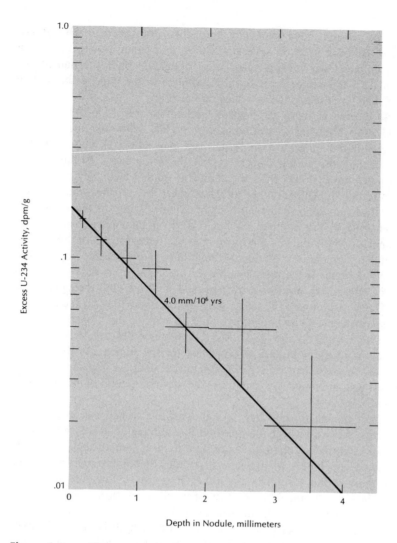

Figure 4-6 Plot versus depth of the natural logarithm of the amount of excess U-234 in the same nodule for which the Th-230 and Pa-231 data were given in Figure 4-5. The slope of the line connecting the points suggests an accumulation rate of about 4 mm/10^6 yrs, as did the other two methods. (Data obtained by T.-L. Ku, Lamont-Doherty Geological Observatory.)

volcanic cores by the potassium–argon method give ages that can be translated to growth rates by dividing the thickness of the Fe–Mn crust by the age of its volcanic center. Rates of between 1 and 3 mm/10^6 yrs have been found in this way.

Finally, the clincher came from a dating method based on a cosmic-ray produced isotope, beryllium-10. (Like C-14, Be-10 is produced by cosmic rays smashing into the atmosphere.) The Be-10 method has been used on several nodules, and again growth rates of a few millimeters per million years were obtained. Since the half-life of Be-10 is 2.5 million years, it could be applied to the bulk of the nodule rather than to only its outermost few millimeters.

So all five of these radioactive methods—three (Th-230, Pa-231, and U-234) applying to the outermost nodule material, one (K–Ar) applying to the innermost material, and one (Be-10) almost wholly encompassing the nodule—seem to tell the same story. Nodules grow 1000 times more slowly than the sediment upon which they rest!

Additional evidence that nodules must be growing very slowly comes from the ratio of the number of nodules lying on the surface in a given area to the number buried in a given thickness of underlying sediment. If Mn nodules were forming at the same rate as the sediments accumulate, mountains of them would have been created in the last millions of years! If the nodules all over the ocean floor today represent only the latest crop of a fast-growing breed, where are all their predecessors? Do they dominate the underlying sediments as they do the surface? Let us consider a hypothetical example. A rectangular trench 4 meters deep is to be dug in the ocean floor. Seven nodules are found on the surface within the boundaries staked out for the trench. As the digging proceeds, the sediment is screened for buried nodules. How many more nodules will be found by the time the desired depth of 4 meters is reached?

Even though no one has ever dug such a trench cut in deep sea sediments, we can answer the question by studying deep sea core collections. Every now and then, the core pipe used to sample such sediments by chance encounters a manganese nodule. Sometimes this encounter takes place at the surface; other times it occurs below the surface in the sediment column. The relative probability of these two types of encounter should give us the answer we seek.

A study of several hundred cores in the Lamont-Doherty Geological Observatory collection for the Pacific Ocean revealed that 54 had encountered manganese nodules. Examination of these cores showed that the number of nodules found buried in the upper 4 meters of the cores was about equal to the number of nodules found at the core tops! In our hypothetical trench, therefore, we would have found only 7 nodules buried in the sediment! Thus if the mining companies who propose to vacuum nodules from the sea floor were to turn to sediment dredging to increase their yield, a column of sediment 4 meters thick would have to be processed to supply a number of nodules equal to those sitting on the surface.

With this observation in mind, we will consider the history of a typical nodule. A shark's tooth falls to the sea floor and begins to accumulate Mn-Fe oxide (see Figure 4-7) at the rate of 4 mm/10^6 yrs. Sediment in the area accumulates at the rate of 2 m/10^6 yrs. Five

Figure 4-7   Shark's teeth that were dredged from the ocean floor in various stages of Mn incrustation. A centimeter ruler is shown to indicate size.

million years pass. The nodule has grown to a radius of 2 centimeters, and 10 meters of sediment have accumulated. (See Figure 4-8 for a cross section of a mature nodule.) At the sediment–water interface, the coated shark's tooth now sits 10 meters higher in elevation than when it fell to the bottom. As the tooth accumulates its oxide coating, it somehow manages to stay atop the growing sediment column. The relative growth rates of sediment and nodule demand that some mechanism be operating on the deep sea floor which rolls the nodule around, preventing it from becoming engulfed in the steady rain of fine sediment. Eventually, however, this nodule will be buried. It may grow so large that whatever mechanism causes it to roll around on the surface will become inadequate and it will be entombed in the falling sediment. Or perhaps a sudden large influx of sediment triggered by a catastrophic earthquake will overwhelm the nodule. A third possibility is that the nodule and its neighbors may grow to such an extent that they run out

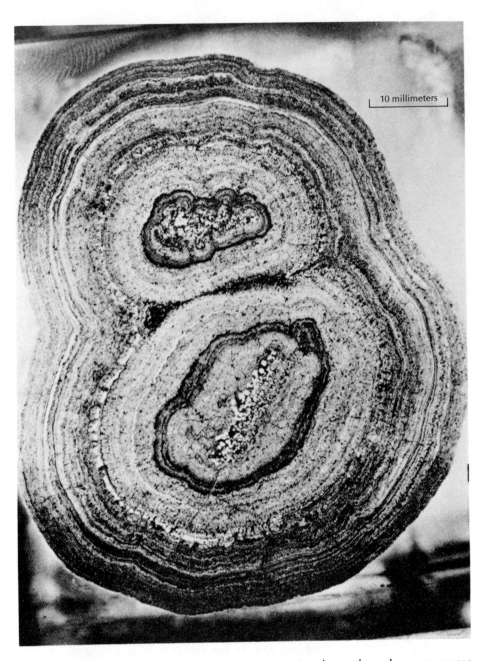

Figure 4-8 Photograph of a cross-sectional cut through a manganese nodule with two growth centers. The altered volcanic material at the centers nucleated the nodule's growth. Crude growth rings representing temporal changes in texture and composition can be seen. This nodule has a radius of 2 centimeters. At a growth rate of 2 mm/10^6 yrs, it must have commenced about 10,000,000 years ago.

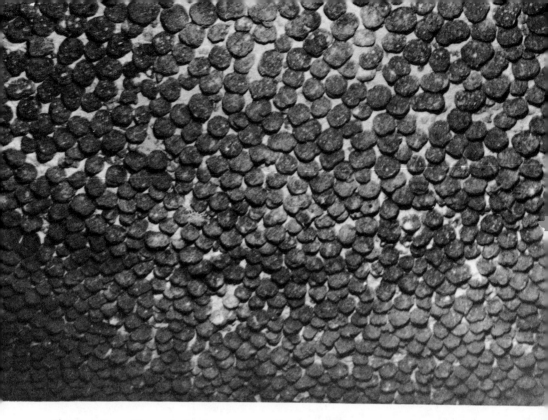

Figure 4-9   Photograph showing a field of unusually closely spaced manganese nodules on the bottom of the Antarctic Ocean. The mean nodule diameter is 6 centimeters.

of room (see Figure 4-9) and become cemented to one another, forming a continuous pavement that is then buried in the sediment.

As far as we know, the life cycle of a manganese nodule consists of growth around a nucleus. Almost every Mn nodule examined has had something at its center—a volcanic fragment, a shark's tooth, a clot of sediment—which is bigger than the fine-grained clay minerals surrounding it. As the nodule grows, it is rolled about and freed of the ever-falling, fine-grained sediment. Whatever keeps the nodule in motion (internal waves in the ocean, tsunami, worms, rat-tailed fish, who knows) allows it to grow for perhaps several million years. And finally, for some unknown reason, the nodule meets its doom; it becomes trapped in the sediment and is buried.

Hence, Mn nodules are so abundant on the sediment surface compared to the underlying sediment because the product of roughly 5,000,000 years of growth is stored at the interface! The nodules from the previous 5,000,000 years are stored in the upper 10 meters of sediment. For each mature nodule that becomes trapped in the sediment, another newly arrived nucleus is beginning its long accumulation of iron and manganese oxide. Like the sharks whose teeth are found at the nodule centers, nodules are born and die; at any given time, nearly the same number inhabit the sea floor.

Manganese Accumulation Rates in Nodules Compared with Red Clays

Knowing the Mn content of nodules and the Mn content of the adjacent sediment on which they rest, we can calculate the rate at which Mn accumulates (per unit area of sea floor) in both the sediment and on the nodules. These rates (shown in Table 4-4) indicate that manganese is accumulating somewhat *less* rapidly in the nodules than in the adjacent sediment! Thus Mn nodules do not grow because of their unusual capacity to attract Mn, Fe, · · ·, but rather because manganese and iron oxides form on all materials exposed to deep sea water. The surrounding sediment is a more effective collector of these oxides than the manganese nodules themselves. Even patches of the ocean floor where basaltic rock has been exposed to the sea for millions of years show about the same accumulation rates: they grow rinds of Mn-Fe oxide about 3 millimeters thick for each million years of exposure!

Table 4-4   Comparison of Mn accumulation rates in nodules and in the sediment upon which they rest. (Data compiled by M. Bender and T.-L. Ku, both of Lamont-Doherty Geological Observatory.)

Ocean	Nodules	Sediment
	Mn, g/cm^2 10^3 yrs	
Whole ocean	.6 (5)*	12. (30)*
North Atlantic	.4 (1)	1.5 (8)
South Atlantic	—	.6 (1)
Indian	—	1.1 (7)
Antarctic	—	.4 (4)
South Pacific	.2 (1)	1.4 (2)
North Pacific	.8 (3)	1.4 (8)

* Numbers in parentheses indicate number of nodules or cores upon which the average is based.

Summary

The mystery of the origin of the manganese nodules that are found scattered over large portions of the ocean floor has been only partially resolved. We know the oxides of Mn and Fe accumulate everywhere on the sea floor. Volcanic rock, sediment, and nodules all receive their share of this material. The mechanism which allows these objects to survive several million years of sediment influx without being buried remains unknown. Continental weathering, upward migration in the sediment column, and submarine volcanism all must contribute to the supply of manganese; what is not clear is which source dominates. We know almost nothing about how these and certain other elements move through the sea because their concentrations are so low that they have defied detailed measurement. The rapid removal of thorium from the sea shows that elements which have a strong preference for particulate phases can be removed extremely rapidly. Perhaps the transitional metals Mn, Ni, Co, Cu, and Fe are analogous to Th in this respect.

SUPPLEMENTARY READINGS

■ General discussions of manganese nodule distributions on the sea floor:

Mero, J. L. *The Mineral Resources of the Sea.* New York: American Elsevier Publishing Co., 1964. Description, distribution, and chemical analyses of manganese nodules found on the deep sea floor.

Heezen, Bruce C., and Hollister, Charles D. *The Face of the Deep.* New York: Oxford University Press, 1971. Photographs of the deep sea floor, many showing manganese nodules.

■ Articles concerning the distribution of Th-228 and Ra-228 in the oceans:

Moore, W. S., and Sackett, W. M. "Uranium and Thorium Series Inequilibrium in Sea Water." *J. Geophys. Res.* 69 (1964):5401–405. Outlines the initial discovery of these isotopes in the sea.

Kaufman, A., Trier, R. M., Broecker, Wallace S., and Feeley, H. W. "The Distribution of Ra-228 in the World Ocean. *J. Geophys. Res.,* 78 (1973): 8827–48. Discusses the distribution of Ra-228 in the surface waters of the world ocean.

Broecker, Wallace S., Kaufman, A., Trier, R. M. "The Residence Time of Thorium in Surface Sea Water and Its Implications Regarding the Rate of Reactive Pollutants." *Earth Planet. Sci. Letters* 20 (1973): 35–44. Discusses the ratio of Th-228 to Ra-228 in various regions of the surface ocean.

■ Publications dealing with sediments rich in iron and manganese:

Degens, E. T., and Ross, D. A. (eds.). *Hot Brines and Recent Heavy Metal Deposits in the Red Sea.* New York: Springer-Verlag New York, Inc., 1969.

A book compiling articles dealing with the metal-rich sediments found beneath pockets of hot salt brine in the deep Red Sea.

Bender, M. L., Broecker, Wallace S., Gornitz, V., Middel, U., Kay, R., Sun, S.-S., and Biscaye, P. E. "Geochemistry of Three Cores from the East Pacific Rise." *Earth Planet. Sci. Letters* 12 (1971):425–33. A discussion of the origin of the metaliferous sediments found on the crest of the East Pacific Rise.

Bender, M. L., Ku, T.-L., and Broecker, Wallace S. "Accumulation Rates of Manganese in Pelagic Sediments and Nodules. *Earth Planet. Sci. Letters* 8 (1970):143–48. Examines the regional variations in the rate of accumulation of manganese in deep sea sediments and in manganese nodules.

■ Papers dealing with the growth rate of manganese nodules:

Barnes, S. S., and Dymond, J. R. "Rates of Accumulation of Ferro-Manganese Nodules." *Nature* 213 (1967):1218. Documentation of the great age of volcanic fragments found in the centers of certain nodules.

Ku, T.-L., and Broecker, Wallace S. "Uranium, Thorium, and Protactinium in a Manganese Nodule." *Earth Planet. Sci. Letters* 2 (1967):317–20. A cross-comparison of the Th-230–Pa-231 and U-234 methods for establishing nodule growth rate.

Krishnaswami, S., Somayajulu, B. L. K., and Moore, W. S. "Dating of Manganese Nodules Using Beryllium-10." *Ferromanganese Deposits on the Ocean Floor,* ed. David R. Horn. Palisades, N.Y.: Lamont-Doherty Geological Observatory, 1972, pp. 117–22. Considers the Be-10 method of determining the growth rate of manganese nodules.

Bender, M. L., Ku, T.-L., and Broecker, Wallace S. "Manganese Nodules: Their Evolution." *Science* 151 (1966):325–28. A comparative discussion of the frequency of nodule occurrence in the sedimentary column and at the overlying sediment interface.

PROBLEMS

4-1   A Mn nodule grows at a radial rate of $2 \text{ mm}/10^6$ years and is 20 percent manganese by weight. The underlying sediment is accumulating at the rate of .1 cm/10^3 yrs and contains 2000 ppm of manganese. Per unit area (as viewed from above), which is "capturing" more Mn—the nodule or the adjacent sediment?

4-2   If element x has a concentration of 500 ppm in basalt and 4500 ppm in granite, what would you expect its concentration to be in shale? If it is rare in Mn nodules, what would you expect its concentration to be in red clay?

4-3   Mn nodules are found at a frequency of $10/\text{m}^2$ on a given region of the sea floor. They average 8 centimeters in diameter, and each has a shark's tooth at its center. If these nodules grow at the rate of $4 \text{ mm}/10^6$ yrs and if all the teeth in all the live sharks in the overlying surface water end up in nodules, how many sharks would you expect to find living in each square kilometer of surface water? (Sharks live for an average of 20 years and each produces 200 teeth.)

4-4   If radium (a radioactive element with a half-life of 1600 years) has a marine geochemistry identical to that of barium (that is, if the Ra/Ba ratio in a marine organism or absorbed on a particle is the same as the Ra/Ba ratio in local water), will more of the Ra atoms added to the sea have fate (a) or fate (b) below?

(a) Radioactive conversion to Rn-222 (the daughter product of Ra-226) while still dissolved in the sea.

(b) Loss to the sediment and subsequent radioactive transformation.

Explain your choice.

4-5   What is the age of the nodules shown in Figure 4-8?

5

ATMOSPHERIC AND

VOLCANIC GASES

Overlying and interacting with the surface ocean is the atmosphere. In our discussion in Chapter 1 of the chemical inhomogeneities within the ocean, we did not treat the dissolved gases because their distribution is influenced not only by the formation and the destruction of particulate material but also by their interchanges with the atmosphere. One of these gases carrying a radioisotope of carbon, $C\text{-}14O_2$, has already entered our story. It is now time to introduce the others.

The elements listed in Table 1-1 include a series of six that are chemically inert—helium (He), neon (Ne), argon (Ar), krypton (Kr), xenon (Xe), and radon (Rn)—all of which exist as gases both in the atmosphere and dissolved in sea water. In addition to these elements, three other dissolved gases—nitrogen, oxygen, and carbon dioxide—are of particular interest to us here. Nitrogen (N_2), although it is converted to and from nitrate ion (NO_3^-) by various organisms living in soils and in natural waters, is much like the six inert gases. It does not undergo any chemical reactions which measurably alter its abundance.* On the other hand, carbon dioxide and molecular oxygen are, as we already know, heavily involved with the chemical processes taking place within

---

* In the deep Pacific, the concentration of the element nitrogen as dissolved N_2 is 1.2 moles/m³, while its concentration as NO_3^- is .045 moles/liter, or 20 times less.

the sea. The distribution of these two gases will primarily occupy our attention.

Oceanographers have long known that oxygen gas is, as the result of consumption by animals and bacteria, deficient in most subsurface ocean waters. The longer a water mass is isolated from the atmosphere, the lower its oxygen content becomes. This fact has enabled O_2 to be widely used as a circulation tracer.

Carbon dioxide (CO_2) has a characteristic not shared by the other gases. As we have seen, most of the dissolved carbon in the sea is in the form of bicarbonate and carbonate ions. In surface water, on the average about one carbon atom in 200 is in the form of CO_2 gas. The interaction among CO_2, HCO_3^-, and CO_3^{--} will prove important in the discussions which follow. As we shall see, the ocean ultimately determines the atmosphere's CO_2 content, rather than vice versa.*

At the surface of the ocean, the amount of all the dissolved gases (except radon)† is controlled by their concentrations in the overlying atmosphere and by temperature. The problem of determining variations in surface ocean concentrations is simplified by the fact that the atmosphere is almost perfectly mixed with regard to the gases we will examine here. Even for CO_2 (the most likely to vary), the difference in the atmospheric concentration between the polar and the equatorial regions is just barely measurable. The assumption that the air over every region of the ocean has the same composition proves quite convenient for our purposes. The total pressure the atmosphere exerts on the sea surface is also very nearly the same from place to place. It varies only because of differences in the H_2O vapor content and temperature (the highs and lows of the meteorologist). The pressures of the gases of interest to us are constant to about ±2 percent at the sea surface.

All deep water was once at the sea surface. When water loses contact with the atmosphere, it carries with it a near equilibrium amount of each atmospheric gas. So when we examine deep sea water, we would expect the content of its gases, except CO_2 and O_2 (and Rn), to be determined by their atmospheric content and by the temperature of the deep water when it left the surface. This has been checked and, indeed, with the exception of helium, the amount of each of these gases in deep water proved to be within a few percents of the expected amount. The deep waters bear witness to the fact that they were once equilibrated with air. The geothermal heat coming up from the sediment and the heat diffusing from the overlying warm surface ocean only slightly change the temperature of the subsurface water masses. Their

---

* Since over 95 percent of the total of all other gases (except radon) resides in the atmosphere, the atmosphere dictates the ocean's gas content.

† Radon is radioactive and has a half-life of only four days. In Chapter 6, we will see that its distribution in the surface ocean is of great importance to our understanding of the flow of gases between the air and sea.

in situ temperatures (when corrected for adiabatic heating)* are fairly close to the temperatures these masses had when they left the sea surface.

Solubilities of Gases in Sea Water

Before we examine in more detail the factors creating anomalies in the He, CO_2, and O_2 (and Rn) content of the oceans, let us first consider how the solubilities of all the gases we will discuss here vary with molecular weight and temperature. The data in Table 5-1 clearly portray the strong correlation between the solubility of a gas and its molecular weight (CO_2 is the marked exception). Note that the higher the mass of the gas molecule the more soluble it is in sea water. Helium (mass = 4) is the least soluble; xenon (mass = 131) is the most soluble of those listed.

In the third column of the table, the solubilities are given as the number of standard cc's of gas† which would dissolve at 0° C and at

Table 5-1 Solubilities of various gases in sea water per atmosphere of gas pressure.

| Gas | Molecular Weight, g/mole | Solubility, cc/liter | | Solubility, moles/m³ | | Solubility 0° C | Air, cc/liter |
		0° C	24° C	0° C	24° C	Solubility 24° C	Water, cc/liter 24° C
He	4	8.0	6.9	.36	.31	1.2	145
Ne	20	9.4	8.1	.42	.36	1.2	124
N₂	28	18	12	.80	.54	1.5	83
O₂	32	42	26	1.9	1.1	1.7	38
Ar	40	39	23	1.7	1.0	1.7	44
CO₂	44	1460	720	65	32	2.0	1.4
Kr	84	71	43	3.2	1.9	1.7	23
Xe	131	136	70	6.1	3.1	2.0	14

* The pressure on a parcel of water increases from 1 to 500 atm as it sinks from the surface of the ocean to a water depth of 5000 meters. This pressure increase results in a slight increase in the density of the water, which in turn leads to a rise in temperature. For a 5000 meter descent, this warming is about .5° C. Thermometers sent to the abyss record temperatures higher than the temperature of the water when it began its descent!

† One standard cc of a gas is the amount contained at 1 atm pressure and 24° C temperature in a volume of one cubic centimeter. The units cc and ml are interchangeable.

24° C in 1 liter of sea water at 1 atmosphere pressure of that gas. For example, if we were to expose sea water at 0° C to pure helium gas at a pressure of 1 atm, 8 standard cc's of He would dissolve in each liter of water; for 1 atm of xenon, 136 standard cc's would dissolve per liter. Xenon is thus about 16 times more soluble than helium.

In the last column of Table 5-1 the solubilities are expressed as the ratio of the number of cc's of each gas per liter of air divided by the number of cc's of that gas per liter of water at equilibrium with the water. These entries might be termed the partition coefficients for the gases. Think of the following experiment. One liter of sea water is placed in contact with 1 liter of air. If some xenon were injected into this system, it would distribute itself in such a way that for every atom of Xe that took residence in the water 14 atoms of Xe would reside in the air. If we were to repeat this experiment with helium, then for every He atom found in the water there would be 145 in the air.

All gases become more soluble in water as the temperature is lowered and less soluble as the temperature is raised. Table 5-1 also lists the solubilities of each gas of interest both at 0° C and 24° C and the ratio of these two solubilities. Figure 5-1 gives this information graphically. Note that the temperature dependence of He is small, changing by only 10 percent over the 24°-temperature range, whereas the temperature dependence of Xe is quite large, changing by a factor of 2 over the same temperature interval. Xenon is twice as soluble in water at 0° C as it is in water at 24° C.

Table 5-2 gives the composition of water-saturated air and the corresponding amounts of each gas to be expected in surface water at 0° C and 24° C (obtained by multiplying the fraction of each gas in air times its solubility at 1 atm pressure). As already stated, the measured values correspond closely to these predicted values.

Table 5-2 Solubilities of various gases in surface ocean water.

Gas	Partial Pressure in Dry Air, atm	Equilibrium Concentration in Surface Sea Water, cc/liter	
		0° C	24° C
H_2	5×10^{-7}		
He	5.2×10^{-6}	4.1×10^{-5}	3.4×10^{-5}
Ne	1.8×10^{-5}	1.7×10^{-4}	1.5×10^{-4}
N_2	.781	14	9
O_2	.209	8.8	5.5
Ar	9.3×10^{-3}	.36	.22
CO_2	3.2×10^{-4}	.47	.23
Kr	1.1×10^{-6}	8.1×10^{-5}	4.9×10^{-5}
Xe	8.6×10^{-8}	1.2×10^{-5}	$.6 \times 10^{-5}$

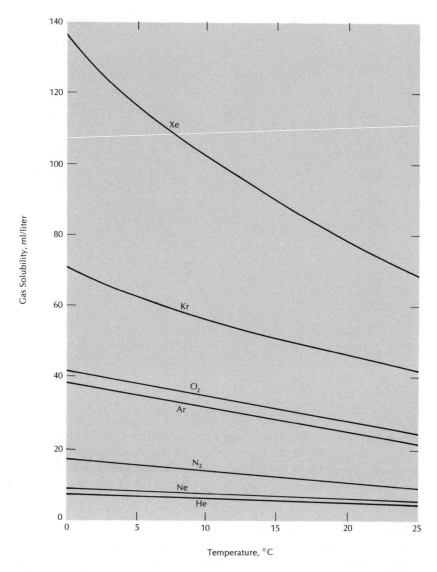

Figure 5-1 The solubilities of various gases in sea water as a function of temperature. The units are standard cubic centimeters of gas contained by a liter of water per atmosphere of pressure exerted by the gas of interest.

Excess Helium

There has been only one great surprise in connection with the gaseous content of sea water. Excesses of helium have been found in deep ocean

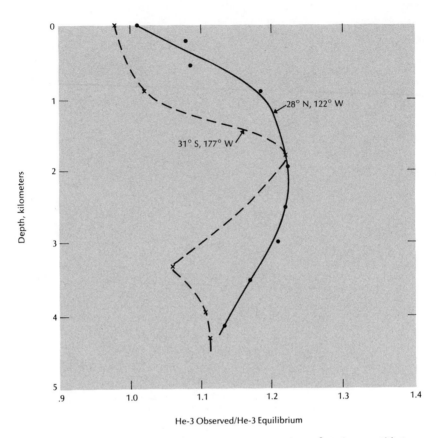

Figure 5-2 Excesses of He-3 over that expected at $0°$ C for equilibrium with the atmosphere at two locations in the Pacific Ocean. (Data based on measurements by W. Brian Clarke and M. A. Beg, McMaster University, on samples collected by H. Craig, Scripps Institution of Oceanography.)

water ranging up to 10 percent more than the deep sea content predicted from the He content of the atmosphere and the temperature of the water! These excesses are thought to be the result of helium leaking from the earth's interior. This is a very exciting discovery because for the first time we will be able to estimate the rate at which the interior of the earth is losing its volatile components. Establishing such a rate will greatly influence our theories of the origin of the earth's atmosphere, ocean, and crust, as we will see in Chapter 7.

Two isotopes of the element helium exist in nature: He-4 with a mass of 4 (2 protons and 2 neutrons), and He-3 with a mass of 3 (2 protons and 1 neutron). The heavier isotope is dominant on the earth's surface: for every million He-4 atoms there is only one He-3 atom. However, the light isotope shows the largest anomaly in sea water: its abundance ranges up to 30 percent higher than the deep sea content predicted from the temperature of the water and the abundance of He-3 in the atmosphere. As shown in Figure 5-2, the greatest He-3 excesses are found about 2.5 kilometers beneath the surface in the Pacific Ocean. This strongly suggests that the excess helium is being added to the ocean from the crests of the mid-ocean ridges which rise to about this elevation. The helium so released appears to spread very rapidly in the horizontal direction and only very slowly in the vertical direction. Once the details of this distribution pattern are known, the mantle-derived He-3 in the ocean will add much to our knowledge of the mixing processes operating within the deep sea.

Recent studies carried out at the Woods Hole Oceanographic Institution reveal that the "plume" of He-3 spreading away from the mid-ocean ridge crests is accompanied by high concentrations of iron in both dissolved and particulate form. If this is true, then part of the great manganese nodule mystery outlined in Chapter 4 may be resolved. The sources for the great excesses of Fe, Mn, and the other metals in deep sea sediment may well be emanations from the ridge crests. Rapid lateral spreading would carry these metals far from their sources. As they traveled, they would coagulate into particles which would gradually settle down onto the underlying sea floor.

Carbon Dioxide Content of Surface Ocean Water

Throughout the ocean, the following chemical reaction is continually taking place:

$$H_2O + CO_2 + CO_3^{--} \leftrightharpoons 2HCO_3^-$$

One water molecule reacts with one dissolved CO_2 gas molecule and with one carbonate ion to form two bicarbonate ions. This reaction proceeds rapidly in sea water and chemical equilibrium is maintained; that is, the concentrations of these ingredients obey the following rule:

$$K = \frac{[HCO_3^-]^2}{[H_2O][CO_2][CO_3^{--}]}$$

where K is the equilibrium constant for the reaction. At any given temperature and pressure, the ratio of the square of the HCO_3^- ion concentration to the product of the concentrations of H_2O, dissolved CO_2, and CO_3^{--} ion must be constant. Since the concentration of the water itself does not change measurably as the result of its participation in the reaction,* the CO_2 content of any sea water sample is proportional to

---

* Sea water contains 54 moles/liter of H_2O molecules and only 2×10^{-3} moles/liter of the various carbon species.

the square of the HCO_3^- ion concentration divided by the CO_3^{--} ion concentration. We have already seen in Chapter 3 that at any place in the ocean the HCO_3^- ion concentration and the CO_3^{--} ion concentrations are fixed by the alkalinity A and the total dissolved carbon content ΣCO_2 of the water. If we write the bicarbonate and carbonate ion concentrations in terms of these quantities, we obtain:

$$[CO_2] = \frac{1}{K} \frac{(2[\Sigma CO_2] - [A])^2}{[A] - [\Sigma CO_2]}$$

The pressure of CO_2 gas in air equilibrated with this water will be:

$$p_{CO_2} = \frac{[CO_2]}{\alpha}$$

where α is the solubility of CO_2 in sea water at the temperature of interest (see Table 5-1). Combining with the previous equation, we then obtain:

$$p_{CO_2} = \frac{1}{\alpha K} \frac{(2[\Sigma CO_2] - [A])^2}{[A] - [\Sigma CO_2]}$$

Since both α and K depend on temperature, so does the equilibrium CO_2 partial pressure p_{CO_2}. A plot of α, K, and the product αK as a function of temperature is given in Figure 5-3.

The temperature dependence of the equilibrium constant product raises a problem. We have implied that most surface water has lost nearly all of its phosphate and nitrate to falling particulate debris and that this brings the alkalinity and total dissolved CO_2 content to nearly constant values. Thus we would predict that the $[HCO_3^-]^2/[CO_3^{--}]$ ratio would be nearly constant throughout the surface ocean. However, since the product αK varies with temperature, the equilibrium CO_2 pressure for any given area of the ocean is dependent on temperature (the equilibrium CO_2 pressure rises 4 percent per degree-centigrade increase in temperature). The equilibrium CO_2 pressure for tropical water at $24°$ C would then be roughly 3 times as great as it would be for high latitude water at $2°$ C! This would tend to create a poleward decrease in atmospheric CO_2 content, but we have already said that no such gradient exists! The atmosphere mixes so rapidly that such a gradient cannot be generated.

If, then, the CO_2 content of high latitude surface sea water is lower and the CO_2 content of tropical water is higher than the CO_2 equilibrium predicted from the well-mixed atmosphere, CO_2 must flow out of tropical waters into the air and from the air into high latitude water. If these CO_2 exchanges between air and water were to reach equilibrium (before the surface temperature was changed by currents or seasons), then the CO_2 excess in low latitude water and the CO_2 deficiency in high latitude water would be eliminated and all surface water would

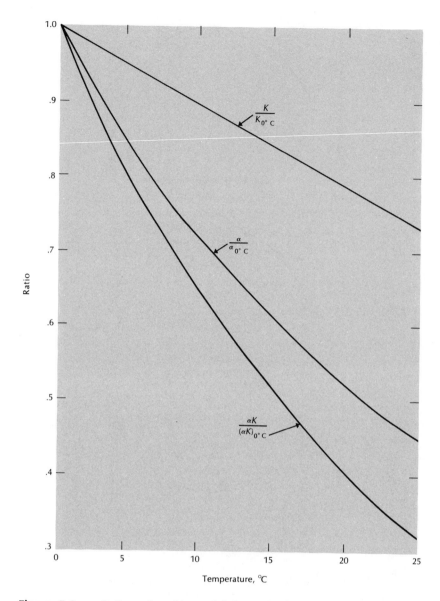

Figure 5-3 Ratios of α (the solubility of CO_2), K (the equilibrium constant for the reaction $H_2O + CO_2 + CO_3^{--} \rightleftharpoons 2HCO_3^-$), and αK at various temperatures to α, K, and αK at $0°$ C. The product αK is inversely proportional to the equilibrium CO_2 partial pressure p_{CO_2} for waters of fixed ΣCO_2 and A. At $0°$ C, the value of α is 65 moles/m^3 atm and the value for K is 1780. Thus, for given values of A and of ΣCO_2, p_{CO_2} will be 1/.324, or 3.1 times greater at $24°$ C than at $0°$ C!

Carbon Dioxide Content of Surface Ocean Water / **123**

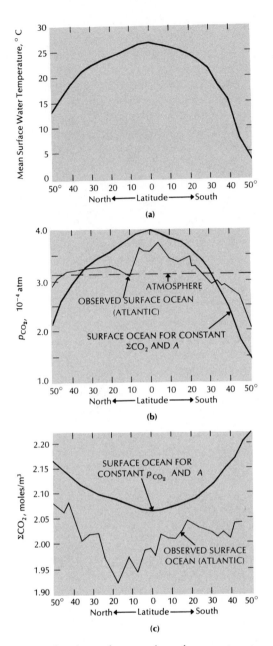

Figure 5-4 Latitude dependence of surface water temperature (a), p_{CO_2} in the air and in the surface ocean (b), and the ΣCO_2 content of surface ocean water (c). For comparison, the p_{CO_2} curve predicted for a uniform surface ocean ΣCO_2 content (2.05 moles/m³) and the ΣCO_2 content curve predicted for a uniform surface ocean p_{CO_2} (3.2 × 10⁻⁴ atm) are shown by the heavy solid lines in (b) and (c), respectively. If the time available for CO_2 equilibration were long compared to the time necessary to achieve equilibrium, the p_{CO_2} curve in (b) would be observed. If, on the other hand, the time available for CO_2 equilibration were short compared to the time necessary to achieve equilibrium, the ΣCO_2 curve in (c) would be observed.

have a CO_2 gas content at equilibrium with the overlying air. The ΣCO_2 would increase toward the poles in accord with the equation:

$$[\Sigma CO_2] = \frac{2[A] - \alpha K p_{CO_2} + \sqrt{(2[A] - \alpha K p_{CO_2})^2 - 16[A](1 - \alpha K p_{CO_2})}}{8}$$

which we obtain by solving the p_{CO_2} equation given previously for ΣCO_2. Although, in this case, all the waters would have the same CO_2 pressure as the air and the same alkalinity, the product αK, and hence the ΣCO_2 content, would depend on the local temperature.

On the other hand, if the time required for equilibration were long compared to the time available, then the ΣCO_2 content would remain nearly constant with latitude. In this case, the dissolved CO_2 gas content of polar water would be lower and the CO_2 content of equatorial water would be higher than the CO_2 content predicted for equilibrium with the atmosphere.

Actually, the situation lies in between these two extremes, as shown in Figure 5-4. Both the ΣCO_2 and the CO_2 gas content of surface ocean water vary with water temperature. This implies that the time constant for equilibration is comparable to the time any given surface water mass remains at a given temperature.

The Rate of Gas Exchange

By making a rather simple model (as we did in the case of vertical mixing in the sea), we can establish the magnitude of the exchange flux of gases between the ocean surface and the atmosphere. As before, we will use the distribution of natural C-14 to "calibrate" this model. Our model assumes that the upper few meters of the ocean have a uniform concentration of the gas of interest. The same assumption is made for the column of air above the sea surface. These two well-mixed reservoirs are separated from one another by a "stagnant" film of water, and gases cross this boundary layer only by *molecular diffusion* (the random motion of individual gas molecules). Although diffusive processes are nondirectional, if a concentration gradient exists, this chaotic motion will lead to a net transfer of material from the zone of high concentration to the zone of low concentration. Thus, if the concentration of gas in the water is not at equilibrium with the concentration of gas in the air, there will be a net flow of gas through the stagnant film. The rate at which the gas is transferred across the air–sea interface then depends upon:

1. the thickness of the film (the thicker the film, the longer the gas molecules will take to wander through it)

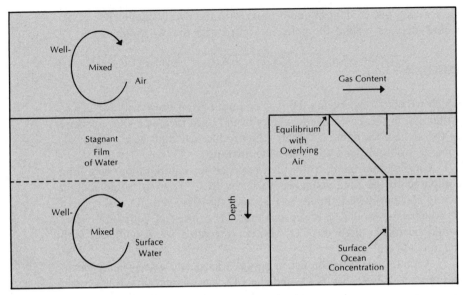

Figure 5-5 Gas exchange model. A thin film of "stagnant" water separates the well-mixed overlying air from the well-mixed underlying surface water. Gases are transferred between air and water only by molecular diffusion through this film. The concentration of gas within the film grades from that corresponding to equilibrium with the overlying air at the top to that found in the surface ocean at the base. The film thickness varies inversely with the degree of agitation of the air–water interface. In the ocean, this film averages several tens of microns in thickness (a micron is one millionth of a meter).

2. the rate at which the gas molecules diffuse through sea water (the warmer the water, the more rapid the molecular motion)
3. the magnitude of the disequilibrium between the gas content of the air and the gas content of the water (the greater the concentration gradient, the more rapid the diffusive transfer)

Figure 5-5 shows the concentration of the gas as a function of depth in ocean water, as depicted by our model. Beneath the base of the stagnant film, the concentration is constant and equal to that in the surface sea water. At the very top of the stagnant film, the concentration is established by the overlying air: the uppermost molecular layers of the film are assumed to be in rapid contact with the air and to have a gas content exactly equal to that dictated by the pressure of the gas in the air and by the air temperature. Between the top and the bottom of the boundary film, diffusion creates a linear gradation from the atmosphere-dictated value to the ocean-dictated value.

If the partial pressure of the gas in the air happened to yield a concentration of gas at the top of the film equal to the concentration below the film, then equilibrium would exist and there would be no net transfer of gas. As much gas would enter the sea as would leave it. If, on the other hand, the gas concentration at the top of the boundary layer was higher or lower than the gas concentration in the water, then a gradient would be established and molecules of gas would be transferred through the stagnant film. The rate of this transfer, or the flux of the gas, is stated as the amount of gas going into or out of a given unit of area of sea surface during a given unit of time. We will state gas fluxes in units of moles of gas transferred across one square meter in one year (moles/m^2 yr).

The flux of any diffusing substance is proportional to the concentration gradient. (The proportionality constant is referred to as the *coefficient of molecular diffusion.*) The concentration gradient is obtained by subtracting the concentration of the gas at the base of the boundary layer from the concentration of the gas at the top of the boundary layer and dividing by the thickness of the layer. Therefore:

$$F = D \, \frac{[\text{gas}]_{\text{top}} - [\text{gas}]_{\text{bottom}}}{z_{\text{film}}}$$

where F is the gas flux, $[\text{gas}]_{\text{top}}$ and $[\text{gas}]_{\text{bottom}}$ are the gas concentrations at the top and the base of the stagnant boundary layer, z_{film} is the thickness of the boundary layer, and D is the coefficient of molecular diffusion. Most ions in solution diffuse at room temperature at a rate of about 1×10^{-5} cm^2/sec. (In our units, this is approximately 3×10^{-2} m^2/yr.) The actual diffusion rates depend on the temperature and the type of gas: D increases with temperature and decreases with the mass of the diffusing molecule (see Table 5-3). The thickness of the boundary film depends on the wind stress; the more turbulent the sea surface, the thinner the film. When the same concentration difference is placed across a thinner film, the resulting concentration gradient

Table 5-3   Rates of molecular diffusion of various gases in water.

Gas	Molecular Weight, g/mole	Diffusion Coefficient, $\times 10^{-5}$ cm^2/sec	
		0° C	24° C
H_2	2	2.0	4.9
He	4	3.0	5.8
Ne	20	1.4	2.8
N_2	28	1.1	2.1
O_2	32	1.2	2.3
Ar	40	.8	1.5
CO_2	44	1.0	1.9
Rn	222	.7	1.4

is larger. Therefore gas flows back and forth between the air and sea more rapidly during periods of high winds than during calm periods.

In the same way that C-14 gave us the value of v_{mix} (the vertical mixing rate), the C-14 distribution between the atmosphere and the surface ocean gives us z (the stagnant boundary layer thickness). We have already mentioned that C-14 atoms are born in the atmosphere, where they become C-14O_2 molecules, and that they are transferred into the ocean by gas exchange. By studying the exchange pattern of this radioactive breed of carbon dioxide, we can find the exchange characteristic of CO_2 itself. If our model is valid, then we can determine the exchange characteristics of the other gases in which we are interested.

We have already said that the C-14 cycle has been running for a long time and that it has reached a steady state; in other words, the amount of C-14 at any point in the system does not change with time. If this is true, then the number of C-14 atoms that are being added to the ocean each year by gas exchange with the atmosphere must exactly balance the number of C-14 atoms that are disappearing within the ocean by radioactive decay. To determine how many C-14 atoms are entering the ocean every year as a result of gas exchange, we multiply the concentration gradient of C-14O_2 across the stagnant boundary layer by the diffusion coefficient for C-14O_2. This can be stated in the form of an equation as:

$$\text{Input of C-14} = D \frac{[\text{C-14O}_2]_{\text{top}} - [\text{C-14O}_2]_{\text{bottom}}}{z_{\text{film}}} A_{\text{ocean}}$$

As mentioned in Chapter 3, C-14 is measured in ratio to normal carbon, or C^*/C. It is therefore appropriate to reexpress the C-14O_2 concentrations as:

$$[\text{C-14O}_2]_{\text{bottom}} = \left(\frac{C^*}{C}\right)_{\substack{\text{surface} \\ \text{ocean}}} [\text{CO}_2]_{\text{bottom}}$$

and

$$[\text{C-14O}_2]_{\text{top}} = \left(\frac{C^*}{C}\right)_{\text{atm}} \frac{\alpha_{\text{C-14O}_2}}{\alpha_{\text{CO}_2}} [\text{CO}_2]_{\text{top}}$$

where $[\text{CO}_2]_{\text{bottom}}$ and $[\text{CO}_2]_{\text{top}}$ are the CO_2 gas concentrations at the base and the top of the stagnant boundary layer, and the ratio $\alpha_{\text{C-14O}_2}/\alpha_{\text{CO}_2}$ reflects the fact that because of their greater mass C-14O_2 molecules (molecular weight = 46) are 1.5 percent more soluble than ordinary CO_2 molecules (molecular weight = 44). Thus the C-14/C ratio at the top of the boundary film is slightly greater (1.5 percent) than the C-14/C ratio measured in atmospheric carbon.

Since there can be no net transfer of ordinary carbon between the ocean and atmosphere (because the system is assumed to be at steady state), then:

$$[\text{CO}_2]_{\text{top}} = [\text{CO}_2]_{\text{bottom}} = [\text{CO}_2]_{\substack{\text{surface} \\ \text{ocean}}}$$

If so, our expression for the input of C-14 atoms can be rewritten as:

$$\text{Input of C-14} = \frac{D[CO_2]_{\substack{\text{surface} \\ \text{ocean}}} A_{\text{ocean}}}{z_{\text{film}}} \left(\frac{\alpha_{C-14O_2}}{\alpha_{CO_2}} \left(\frac{C^*}{C}\right)_{\text{atm}} - \left(\frac{C^*}{C}\right)_{\substack{\text{surface} \\ \text{ocean}}} \right)$$

The number of C-14 atoms being lost by radiodecay must match this gain. To find an expression for the loss rate per year, we begin with the number of C-14 atoms in the sea. To calculate this, we multiply the volume of the ocean (expressed as the area of the surface ocean A_{ocean} times the mean depth of the ocean $\hbar$) by the average total dissolved carbon content (mainly HCO_3^- and CO_3^{--} ions) and then by the average ratio of C-14/C in this oceanic carbon. We then multiply this number of C-14 atoms in the sea by the fraction of them undergoing radioactive decay per year λ. This yields:

$$\text{Loss of C-14} = A_{\text{ocean}}\hbar \, [\Sigma CO_2]_{\text{ocean}} \left(\frac{C^*}{C}\right)_{\text{ocean}} \lambda$$

Equating the preceding input expression with this one for loss and solving for z_{film}, we obtain:

$$z_{\text{film}} = \frac{D}{\lambda \hbar} \frac{[CO_2]_{\text{surface}}}{[\Sigma CO_2]_{\text{ocean}}} \frac{\left(1 - \dfrac{(C^*/C)_{\text{surface}}}{(C^*/C)_{\text{atm}}} \dfrac{\alpha_{C-14O_2}}{\alpha_{CO_2}}\right)}{\dfrac{(C^*/C)_{\text{ocean}}}{(C^*/C)_{\text{atm}}} \dfrac{\alpha_{C-14O_2}}{\alpha_{CO_2}}}$$

We have already seen that the average value for the molecular diffusion coefficient D is 3×10^{-2} m²/yr for CO_2 in warm sea water, that the mean depth of the ocean $\hbar$ is 3800 meters, and that the reciprocal of the decay constant of C-14 $1/\lambda$ is 8200 years. The average concentration of CO_2 gas in the surface ocean is .01 moles/m³, the average ΣCO_2 content of ocean water is 2.4 moles/m³, and the $\alpha_{C-14O_2}/\alpha_{CO_2}$ ratio is 1.015. The C-14/C ratio in surface ocean water is, on the average, 3.5 percent lower than the C-14/C ratio in the atmosphere (see Figure 3-4). This last factor is the one that leads to the largest uncertainty in the determination of z_{film}. If only a 1 percent error existed in the measurement of the average C-14/C ratio in surface water carbon relative to that in atmospheric carbon, it would create a 25 percent error in the estimate of z_{film}!

When z_{film} is calculated from these figures we obtain 1.7×10^{-5} meters, or 1.7×10^{-3} centimeters. For lengths this small, it is more convenient to use the unit micron (one millionth of a meter). So z_{film} is 17 microns. Keep in mind that the boundary is thinner than this average value in areas of high wind velocity and thicker in areas of low wind velocity; our model gives us only an oceanwide average.

Since the coefficients of molecular diffusion do not vary greatly from gas to gas, it proves convenient to think of the gas exchange process in terms of the ratio D/z. Since the dimensions of D and z_{film}

are m^2/yr and of meters, respectively, the ratio D/z has the dimensions of m/yr, or of velocity. We might think of this ratio as the velocity of two imaginary pistons: one moving up through the water pushing ahead of it a column of gas with the concentration of the gas in surface water, and one moving down into the sea carrying a column of gas with the concentration of the gas in the upper few molecular layers of the water column. For the average surface ocean, these pistons move at the rate of 3×10^{-2} m^2/yr divided by 1.7×10^{-5} meters, or 1700 m/yr. Thus each day a column of sea water between 4 and 5 meters thick exchanges its gas with the atmosphere! The average gas molecule resides in the upper 100 meters of the ocean between 20 and 25 days before returning to the atmosphere.

The piston concept clarifies the fact that even if the atmosphere and ocean are at equilibrium, the transfer of gas continues: the amount "pushed in" just matches the amount "pushed out."

Rate of ΣCO_2 Equilibration between the Surface Ocean and the Atmosphere

We can apply the piston concept to the question of the extent of ΣCO_2 equilibration between the atmosphere and the local sea surface. The partial pressure of CO_2 in the atmosphere is 3.2×10^{-4} atm. As the solubility of CO_2 gas in sea water at $20°$ C (the average surface temperature) is 33 moles/m^3 atm, the concentration of CO_2 being "pushed into" the sea by the ingoing piston is 1×10^{-2} moles/m^3. At the average piston velocity of 1700 m/yr, 17 moles of CO_2 enter each square meter of sea surface each year.

With this fact in mind, let us consider the response of a sea water sample initially at equilibrium with the atmosphere at $20°$ C and then quickly cooled to $19°$ C. How much time will be required for its ΣCO_2 content to adjust to the new temperature? Because of the shift in the value of the equilibrium constant, such a cooling will lower the CO_2 gas content of the water by 4 percent (that is, 4 percent of the CO_2 present will combine with CO_3^{--} ions to form HCO_3^- ions). From our earlier equation (page 125) relating the ΣCO_2 and the CO_2 ion content of sea water with fixed A, we can show that in order to reestablish equilibrium with the overlying air the ΣCO_2 of the sea water sample will have to increase by .4 percent, or by 8×10^{-3} moles/m^3. Since, after cooling, the CO_2 gas content at the base of the boundary layer is 4 percent lower than that dictated by the atmosphere, the invasion rate of CO_2 should be 4 percent higher than its evasion rate. Thus the net influx of CO_2 will be $.04 \times 17$ or .7 moles/m^2 yr. If the thickness of the well-mixed surface water layer is 100 meters, then the ΣCO_2 deficiency beneath each square meter of surface water immediately after cooling will be 100 meters $\times\, 8 \times 10^{-3}$ moles/m^3, or .8 moles/m^2. The time required for an amount of CO_2 gas equal to the ΣCO_2 deficiency in the 100-meter water column to invade the surface ocean will be just over one year (.8 moles/m^3 divided by .7 moles/m^3 yr, or 1.1 years). Of course, as the ΣCO_2 of the water rises, the CO_2 gas content difference across the boundary film falls. The net rate of CO_2 addition to the water column will fall correspondingly. The ΣCO_2 in the water will change in the same logarithmic manner characteristic of radioisotopic decay. Half the initial deficiency will be eliminated in 1.1 $\times$ .693 or .8 years. Half of the remaining deficiency will be removed in the next .8 years, and so forth. These relationships are shown graphically in Figure 5-6.

Seasonal changes in ocean temperature occur, of course, on a time scale of six months. The mixing of waters in the equatorial ocean with those in the subpolar regions occurs on a time scale of a few months to a few years. Thus equilibration between surface water and the air in any given locality never quite reaches completion. The time constants for ΣCO_2 equilibration and for temperature change are similar. Most, but not all, of the adjustment in total dissolved carbon content takes place.

The Oxygen Cycle in the Surface Sea

As C-14O$_2$ and CO_2 share the same chemistry, there is no reason to doubt the validity of the CO_2 exchange rate calculated from the distribution of C-14. Any inadequacies of the model cancel out when transferring from one isotope to the other. When applied to O_2, however, more careful consideration has to be given to the validity of the model. Specifically, the chemical reaction between CO_2 molecules and the ionic carbon forms provides a complication which must be taken into account.

Two HCO_3^- ions can react to yield a CO_3^{--} ion and a CO_2 molecule. This provides an alternate route for the movement of CO_2 through the boundary layer. The reaction between HCO_3^- ions in the upper regions of the film can supply CO_2 for escape to the atmosphere. Conversely, the combination of invading CO_2 with CO_3^{--} ions in the film can speed the uptake of CO_2 by the solution. As our model assumed no such possibilities, the value of z_{film} we obtained from the C-14 distribution could be too small. In this case, applying z_{film} to oxygen would yield too high an exchange rate. Fortunately, the alternate route is of secondary importance in this situation because the error introduced by its neglect is quite small. We can with reasonable assurance apply the stagnant boundary film thickness obtained for C-14O$_2$ to O_2.

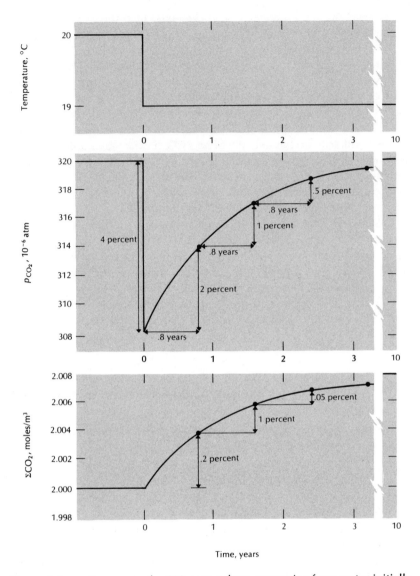

Figure 5-6   Response of a 100-meter deep reservoir of sea water initially at equilibrium with the air to a sudden cooling from 20° C to 19° C. Its p_{CO_2} drops by 4 percent (12×10^{-6} atm). The invasion of atmospheric CO_2 then causes the ΣCO_2 content to rise and the p_{CO_2} to move back toward the atmospheric value (320×10^{-6} atm). The half-time for the reestablishment of air–water equilibrium is about .8 years.

Sea water at equilibrium with the air contains, on the average, .3 moles/m^3 of O_2. Thus the well-mixed upper 100 meters of the sea contain 30 moles/m^2 of O_2. At a piston velocity of 2000 m/yr, about 600 moles of O_2 will be exchanged across each square meter of interface in a year. Thus O_2 atoms reside in the surface ocean only about six weeks before returning to the atmosphere.

This is why surface ocean water never has an O_2 excess (due to plant production) or an O_2 deficiency (due to animal uptake). The rate of O_2 exchange is far more rapid than the processes taking place within the ocean that release or consume oxygen. Measurements of the primary productivity of plants in the ocean show that each year about 12 moles/m^2 yr of O_2 are released by the formation of plant material. (As we will see in a moment, for every unit of carbon fixed by plants 1.4 units of oxygen are produced.) Even if animals dwelling in the surface ocean did not reuse most of this O_2, its pressure in surface water would, on the average, have to be only 1.02 times that in the atmosphere to dispatch this excess oxygen to the atmosphere as fast as it is manufactured. The 2 percent excess O_2 content would raise the evasion rate by .02 × 600 or 12 moles/m^2 yr. As most of the organic matter formed in any given year is oxidized within the surface water, animals must be removing most of the O_2 so formed. We saw in Chapter 3 that only about one-sixth of the carbon produced is not consumed by surface-dwelling animals. Of the 12 moles/m^2 yr of O_2 generated by plants, an average of 10 are used by animals (and bacteria) and surface water has to dispense only 2 to the atmosphere. Thus 602 moles of O_2 evade and 600 moles of O_2 invade the average square meter of sea surface each year. To create this excess outward flux, the O_2 content of surface water need average only .3 percent higher than the O_2 content necessary for equilibrium within the atmosphere (see Figure 5-7).

Actually, measurements show that the surface ocean has a consistent O_2 supersaturation of about 5 percent (see Figure 5-8). Since similar supersaturations are also found for Ar and for N_2, the O_2 supersaturation cannot be attributed to plant productivity. The excess O_2 may result from the entrainment of air bubbles created at the crests of waves and churned down into the water column where part of the gas they contain is forced into solution.

Oxygen Deficiencies in the Deep Sea

Water sinking to the deeps would thus be expected to bear a near equilibrium amount of O_2. Once at depth, the oxygen consumed by animals and bacteria can no longer be replaced by atmospheric exchange or plant growth. Can the O_2 decrease which ensues be simply related to the consequent carbon, nitrate, and phosphate increases?

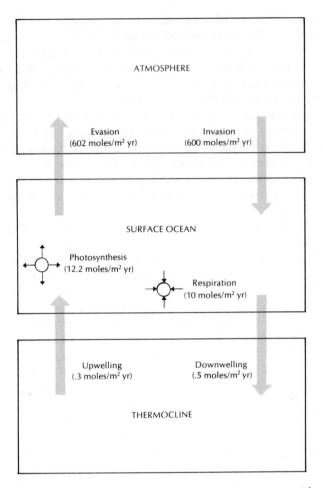

Figure 5-7   The oxygen cycle in the warm surface ocean. Three processes are at work here: exchange with the atmosphere, exchange with the underlying water, and the activity of organisms within the surface ocean. Atmospheric exchange is so rapid compared to the other two processes that little deviation from equilibrium with the atmosphere is expected (that is, the partial pressure of oxygen in the warm surface ocean is 602/600 times that in the atmosphere).

The ratio between carbon release and oxygen consumption during oxidation of organic material has been established in the laboratory. When the debris of marine plankton are burned, about 1.4 moles of oxygen gas are required to combine with each mole of organic carbon present in the sample. In other words, in this experiment, for each 1.0 mole of CO_2 that appears during combustion 1.4 moles of O_2 gas dis-

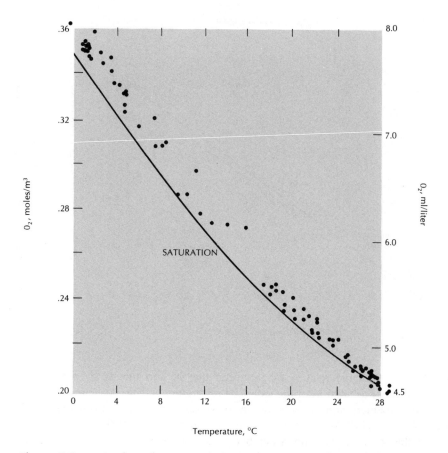

Figure 5-8 A plot of measured O_2 concentrations for surface water samples of varying temperatures from throughout the Atlantic Ocean. (Data obtained by Arnold Bainbridge and his coworkers in the GEOSECS expeditions.) For comparison, the O_2 saturation curve as a function of temperature is given.

appear. In the reaction, oxygen combines with the organic material and the combustion products are CO_2, NO_3^-, and H_2O. The chemical reactions are:

$$O_2 + CH_2O \rightarrow CO_2 + H_2O$$
$$2O_2 + NH_3 + CO_3^{--} \rightarrow NO_3^- + HCO_3^- + H_2O$$

where CH_2O and NH_3 are the major chemical forms of carbon and nitrogen, respectively, in organic tissue. About one O_2 molecule is required for each C atom present in the organic tissue, and 2 O_2 molecules are required for each N atom present. The other constituents

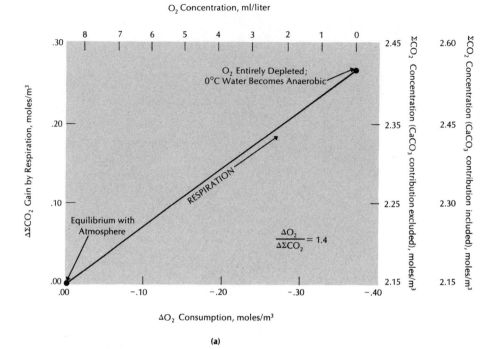

O$_2$ Concentration, ml/liter

O$_2$ Entirely Depleted;
0°C Water Becomes Anaerobic

RESPIRATION

$$\frac{\Delta O_2}{\Delta\Sigma CO_2} = 1.4$$

Equilibrium with
Atmosphere

ΔΣCO$_2$ Gain by Respiration, moles/m^3

ΣCO$_2$ Concentration (CaCO$_3$ contribution excluded), moles/m^3

ΣCO$_2$ Concentration (CaCO$_3$ contribution included), moles/m^3

ΔO$_2$ Consumption, moles/m^3

(a)

Figure 5-9   Predicted relationship between the O$_2$ and the ΣCO$_2$ contents of deep sea water as a function of the extent of respiratory activity (a), compared with actual data obtained from a wide variety of subsurface samples from the Pacific Ocean (b). The right-hand scales for (a) apply only to waters at 0° C, as does the point on this graph where the water becomes anaerobic. The ΔO$_2$ values were calculated by subtracting the measured O$_2$ contents from the O$_2$ saturation content. The ΔΣCO$_2$ values were calculated as follows. First corrections were made for salinity and alkalinity differences, yielding values that would be observed were the ocean uniform in salinity and free of CaCO$_3$ production and dissolution effects. Then the ΣCO$_2$ contents corresponding to zero O$_2$ utilization were subtracted from these values. These differences reflect changes in ΣCO$_2$ content resulting from the oxidation of organic tissue. (Data collected by Charles Culberson, Oregon State University.)

which undergo oxidation are so low in abundance relative to C and N that their O$_2$ requirements can be neglected. Since there are about 2 atoms of N for every 10 atoms of C in falling organic debris, 14 molecules of O$_2$ are required for each 10 atoms of C.

Presumably we could predict how much oxygen has been used in each unit of volume of North Pacific Deep Water merely by determining its amount of excess carbon. Water descending into the deep North

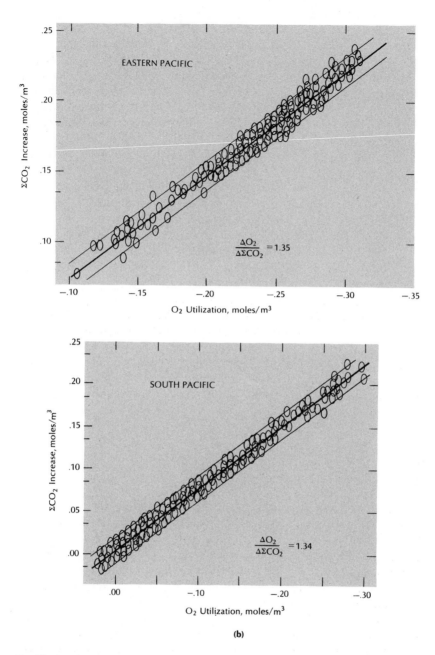

EASTERN PACIFIC

$\dfrac{\Delta O_2}{\Delta \Sigma CO_2} = 1.35$

ΣCO_2 Increase, moles/m³

O_2 Utilization, moles/m³

SOUTH PACIFIC

$\dfrac{\Delta O_2}{\Delta \Sigma CO_2} = 1.34$

ΣCO_2 Increase, moles/m³

O_2 Utilization, moles/m³

(b)

Pacific begins its journey at the surface of the ocean carrying an amount of CO_2 typical of the cold surface ocean: about 2.15 moles of ΣCO_2 per cubic meter of sea water. When this water arrives in the deep North Pacific, it has inherited an extra .30 moles of C. From the alkalinity change, it is possible to calculate that about .10 moles of this carbon is the result of $CaCO_3$ dissolution. Thus .20 moles/m^3 of C must be of organic origin. The oxygen used by animals to liberate that amount of CO_2 from organic material would be $1.4 \times .20$ or .28 moles/m^3 of O_2.

Now let us compare this predicted value with the actual observed deficiency of oxygen in North Pacific Deep Water. When water leaves the surface of the earth in cold regions, it carries an equilibrium amount of oxygen, or 8 ml/liter. From this, we subtract the average observed O_2 content in deep North Pacific water (2 ml/liter). The difference of 6 ml/liter between these two oxygen contents is then a measure of the oxygen consumption.

At this point, we must convert the units used by geochemists (moles/m^3) to the units used by oceanographers (ml/liter). There are 1000 liters in a cubic meter; 22,400 ml of gas (at standard conditions) make up a mole. So to convert from moles/m^3 to ml/liter, we multiply by 22,400 and divide by 1000. Thus the .28 moles/m^3 of O_2 needed for oxidation equals 6.27 ml/liter of gas. The oxygen deficiency of 6 ml/liter can be accounted for by the carbon excess!

Another way to demonstrate this equivalence is to plot the O_2 deficiency in any given water sample against its ΣCO_2 excess (corrected for $CaCO_3$). If this is done for water samples from a wide variety of depths and locations in the ocean, they should fall along a straight line with a slope of 1.4 moles of O_2 per mole of C. Such a plot showing the predicted relationship appears in Figure 5-9(a); results from actual deep Pacific samples are shown for comparison in Figure 5-9(b).

Summary

In this chapter, we have considered the processes controlling the concentrations of dissolved gases within the sea. We have seen that Ne, Ar, Kr, Xe, and N_2 exist in the sea in nearly the same concentrations expected if the water had once equilibrated with the atmosphere at its *in situ* temperature. Helium (especially its light isotope He-3) shows an excess in deep water thought to be the result of the injection of gas from the earth's interior at the crests of the mid-ocean ridges. The concentration of CO_2 gas in surface water reflects that in the overlying air. Because waters with the same alkalinity and different temperatures have, at equilibrium with the well-mixed atmosphere, different ΣCO_2 contents, there is a net evasion of CO_2 from the surface sea into the equatorial atmosphere and a net invasion from the atmosphere into the polar oceans. Deep water is deficient in O_2 content from that predicted for equilibrium with the atmosphere at its *in situ* temperature by amounts in accord with the excess carbon generated by the oxidation of organic material within the deep sea. From the distribution of C-14O$_2$ between the ocean and atmosphere, it is possible to estimate the rates of transfer across the air–sea interface for the gases of interest.

SUPPLEMENTARY READINGS

■ Textbooks dealing with the physical chemistry of gases:

Moore, Walter J. *Physical Chemistry,* Second Edition. Englewood Cliffs, N.J.: Prentice-Hall, 1955. A textbook on classical physical chemistry covering the basic laws of gas interactions.

Danckwerts, P. V. *Gas-Liquid Reactions.* New York: McGraw-Hill, 1970. A chemical engineering textbook concerned with the kinetics of gas exchange.

■ Articles dealing with gas exchange between the ocean and the atmosphere:

Craig, H., Weiss, R. F., and Clarke, W. B. "Dissolved Gases in the Equatorial and South Pacific Ocean. *J. Geophys. Res.* 72 (1967):6165. Paper discussing the distribution of dissolved gases in the sea and giving possible reasons for the observed supersaturation.

Craig, H., and R. F. Weiss, "Dissolved Gas Saturation Anomalies and Excess Helium in the Ocean," *Earth Planet. Sci. Letters* 10, (1971): 289–96. Another paper discussing the distribution of dissolved gases in the sea and giving possible reasons for the observed supersaturation.

Bieri, R. H., "Dissolved Noble Gases in Marine Waters," *Earth Planet. Sci. Letters* 10, (1971): 329–33. A third paper discussing the distribution of dissolved gases in the sea and giving possible reasons for the observed supersaturation.

Bolin, B. "On the Exchange of Carbon Dioxide between the Atmosphere and the Sea." *Tellus* 12 (1960):274. A theory of the exchange of CO$_2$ gas between the ocean and atmosphere.

Broecker, Wallace S., and Peng, T.-H. "Gas Exchange Rates between Air and Sea." *Tellus,* 1974. A review of the measurements of gas exchange rate between the ocean and atmosphere.

■ Publications dealing with the distribution of dissolved inorganic carbon and dissolved oxygen gas within the sea:

Riley, J. P., and Skirrow, G. (eds.). *Chemical Oceanography,* Volumes 1 and 2. London and New York: Academic Press, 1965. A general discussion of the distribution of HCO$_3^-$, CO$_3^{--}$, and CO$_2$ giving the equations needed to precisely calculate the proportionation of these properties as a function of the ΣCO_2 and A of a water sample (including the influence of boron).

Culberson, C., and Pytkowicz, R. M. "Oxygen–Total Carbon Dioxide Correlation in the Eastern Pacific Ocean." *J. Oceanographic Society of Japan* 26 (1970):95–100. Discusses the correlation between O_2 consumption and CO_2 production in subsurface waters.

■ Papers dealing with the distribution of the noble gases in the sea:

Clarke, W. B., Beg, M. A., and Craig H. "Excess Helium-3 at the North Pacific GEOSECS Station." *J. Geophys. Res.* 75 (1970): 7676–78. The discovery of excess He-3 in the sea.

Bieri, R. H., Koide, M., and Goldberg, E. D. "Noble Gas Content of Marine Waters." *Earth Planet. Sci. Letters* 4 (1968):329. Measurements of the gases He, Ne, Ar, Kr, and Xe throughout the sea.

PROBLEMS

5-1   A tank of gas-free water is exposed to the air. Its temperature is 2° C, and the depth of the tank is 3 meters. After an hour, the O_2 concentration is found to be 10 percent that expected for equilibrium with the atmosphere. What is the stagnant boundary layer thickness for the water in this tank?

5-2   A sea water sample is placed in the hot sun and its temperature rises from 15° C to 25° C. If it initially had a p_{CO_2} equal to that in the atmosphere (3.2×10^{-4} atm), what would the partial pressure of CO_2 gas be once it was heated? The alkalinity of the water is 2.00×10^{-3} moles/liter. (Assume that no biological activity or gas exchange occurs during the heating.)

5-3   Mono Lake in California has a very high ΣCO_2 of .2 moles/liter. The lake has a mean depth of 30 meters and receives new carbon from inflowing streams at a negligible rate. The C-14/C ratio in its dissolved carbon is uniform with depth and is 30 percent lower than that expected if it were at equilibrium with the overlying air. If the C-14 deficiency is at steady state, what must the z value for this lake be? (*Hint:* As in the ocean, the loss of C-14 by radiodecay within the lake is just balanced by the net influx from the air.)

5-4   Helium has a molecular diffusivity three times greater than that for argon. If a tank of water is charged with .5 atm of He and .5 atm of Ar and then allowed to stand open in contact with the air, what percent of the Ar will have been lost when 10 percent of the He has been lost? What is the ratio of the "piston velocities" for the two gases?

5-5 Equal portions of two water samples (one at 5° C and one at 25° C), each initially saturated with the same atmospheric gases, are mixed in such a way that no gas is lost or gained. The concentration of each atmospheric gas in this mixture is then compared with that in a water sample at 15° C and at exact equilibrium with the air. Differences are found. Why? Would the anomaly be larger for helium or for xenon? Why?

6

ISOTOPES AS

WATER MASS TRACERS

We have already touched upon the manner in which radioactive isotopes can yield valuable information about the rates of oceanic processes. We have used the radiocarbon contrast between ocean and atmosphere to determine the rate of gas exchange across the air–sea interface, the radiocarbon contrast between the warm and cold oceans to determine the rate of exchange of water across the main oceanic thermocline, and the distribution of C-14, Th-230, Pa-231, U-234, and Be-10 in sediments and in manganese nodules to determine sedimentation rates. In this chapter we will explore the use of these isotope tracers in studying individual water masses, and we will add some new radioactive isotopes as well as some *stable* or nonradioactive isotopes to our arsenal of tracers.

Water Mass Tracers

One matter of particular interest to physical oceanographers is the origin of the water found in the deep sea. We have already stated that the main source of deep water was the Norwegian Sea and that an auxiliary source existed in the Antarctic. But how do we know this? Physical oceanographers have long used plots of temperature versus salinity to identify water types. However, temperature is a somewhat dangerous tool. Although the subsurface heating of deep water by the diffusion of heat through the thermocline and from the underlying sediments does not greatly alter the temperature of deep water masses, the *in situ* heating is great enough to generate some uncertainties as to the reliability of using temperature–salinity diagrams to establish the relative importance of various possible deep water sources. Oceanographers needed a second property to substitute for temperature in such a diagram.

It was found that the isotope O-18 (a trace species of oxygen which is *not* radioactive) could be used for this purpose. Oxygen-18 was produced long ago, along with ordinary oxygen O-16, in the centers of stars. But because O-18 had a higher nuclear reactivity than O-16, only one atom of O-18 was produced for every 500 atoms of O-16. In the ocean, one out of every 500 water molecules bears an O-18 instead of an O-16 atom.

The ratio of O-18 to O-16 is useful in oceanography because during evaporation and during condensation water molecules containing the heavier oxygen isotope undergo small separations from those containing the lighter oxygen. Waters found at various places on the earth's surface reveal compositions ranging over several percent in their proportions of O-18 and O-16. Water molecules containing O-18 evaporate less rapidly than those containing O-16, so vapor given off by the ocean is depleted in the heavier isotope of oxygen. This depletion is about four-fifths of 1 percent; that is, there are four-fifths of 1 percent less O-18 in vapor leaving the ocean than in the same number of molecules of the surface water supplying the vapor. When this process is reversed and the vapor is converted back to a precipitate as rain or snow, the fractionation is also reversed: water droplets forming in a cloud are enriched in the heavier O-18 relative to the vapor from which they are generated. A likely conclusion might be that since the enrichment of O-18 during condensation just balances its depletion during evaporation, there would be no net effect. The *first* batch of rain to be released by an air mass does have the same O-18/O-16 ratio as the ocean water from which the vapor was derived. Beyond this, however, some cumulative effects lead to important changes in the isotopic composition of the air mass and of the subsequent rain (and snow) that this mass produces.

Before exploring the nature of these changes, we must pause to consider some isotope terminology which will simplify the following discussions. Oceanographers working with these methods express the results of their measurements in terms of a difference between the O-18 content of the sample of water of interest and the O-18 content of a suitable standard. In the case of the O-18/O-16 ratio, the standard is a sea water sample referred to as Standard Mean Ocean Water (SMOW).

Differences in isotopic composition are expressed as per mil* deviations from the isotopic composition of this standard. These per mil differences, called δO-18 values, are calculated from the isotopic compositions of the standard and the sample as follows:

$$\delta\text{O-18} = \frac{(\text{O-18/O-16})_{\text{sample}} - (\text{O-18/O-16})_{\text{SMOW}}}{(\text{O-18/O-16})_{\text{SMOW}}} \times 1000$$

Vapor evaporating from SMOW, for example, would have a δO-18 equal to −8 per mil; that is, it would be .8 percent depleted in O-18 content relative to SMOW.

Now let us return to the reason O-18/O-16 variations are found in surface water. Vapor leaving the sea surface has an average δO-18 value of −8 per mil. The air mass carrying this vapor cools as it rises, and rain forms when the dew point (~20° C) is reached. The first rain drops would, in their formation, undergo exactly the opposite fractionation that occurs during vaporization: *more* H_2O-18 molecules would condense in relation to H_2O-16. The first rain would therefore have an isotopic composition identical to that of the ocean water from which the vapor originated. The remaining vapor would consequently show an even *greater* depletion in O-18. If the air mass were then to move poleward and become cooler, additional rain would form. This rain would have an O-18 content *less* than that in mean sea water! The further poleward the air mass moved, the colder it would become. This would correspondingly reduce its water vapor content and the residual vapor would be ever more depleted in O-18 (since the water falling from the clouds as rain always carries a higher proportion of O-18 relative to O-16 than is present in the cloud vapor). As the amount of moisture held by an air mass decreases in a predictable manner (8 percent per degree centigrade temperature drop), the change in the isotope composition of the precipitation can be readily calculated. Figure 6-1 shows the relationship obtained when this calculation is made.

Most of the water vapor held in clouds comes from the equatorial region of the ocean, which has an average air temperature of 25° C. When cooled to −15° C (a typical winter temperature in northern Canada), there is a hundredfold decrease in water vapor content. So the air masses that arrive in the northern regions of the globe have lost roughly 99 percent of their water, and the residual water vapor in the air has a δO-18 of −45 per mil. A snowfall from this vapor would have a δO-18 of −35 per mil. (As the air cools, the fractionation factor between cloud vapor and precipitation changes a bit; at 0° C it has increased from 8 to 10 per mil.)

---

* As we learned earlier, a per mil is a part per thousand; a percent is a part per hundred.

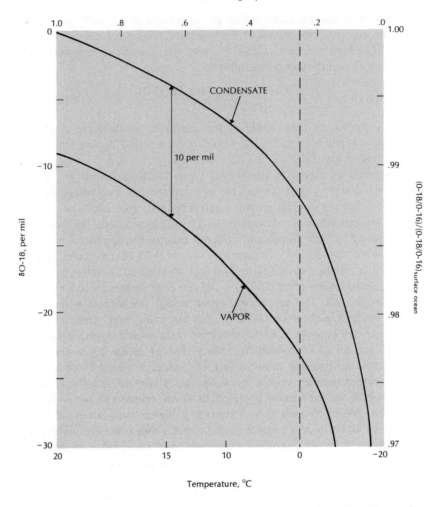

Figure 6-1 Plot of the O-18/O-16 ratio (expressed as δO-18) in the residual cloud vapor and precipitation forming within the cloud versus decreasing cloud temperature (and hence vapor content of air mass). The air mass is assumed to have begun with an O-18/O-16 ratio equal to that for water vapor evaporating from the tropical ocean. When this air reaches its dew point (~20° C), precipitation with an isotopic composition nearly equal to that in the surface ocean (δO-18 = 0) forms. Further cooling reduces the saturation water content, causing the additional precipitation shown on the upper horizontal scale. Since the isotopic composition of the precipitation is always richer in O-18 than the vapor from which it forms, both the δO-18 of the residual vapor and of the precipitation forming from this vapor decrease as the temperature falls. At 0° C, only 25 percent of the original vapor remains. Rain forming from this vapor would be 13 per mil lower in O-18 than surface ocean water. The vapor itself would be about 23 per mil lower in O-18 relative to O-16 than surface ocean water.

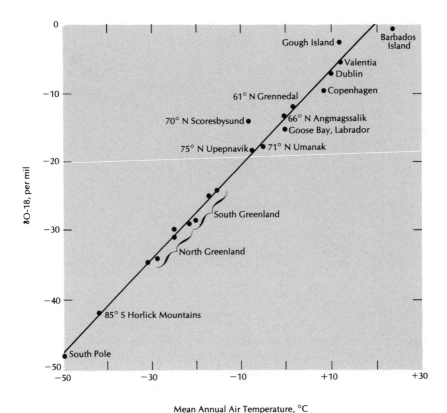

Figure 6-2 Plot of the observed change in δO-18 for average annual precipitation with mean annual air temperature. The sampling locations are indicated. The solid line is the isotopic composition predicted from Figure 6-1. (Data collected by W. Dansgaard, Geophysical Institute, University of Copenhagen.)

Thus precipitation samples collected from various places on the globe should yield an interesting pattern in oxygen isotope composition. (Figure 6-2 gives this data.) Rain falling in the tropical regions has about the same isotopic composition as surface sea water (SMOW); rain falling in Chicago has an average δO-18 of −10 per mil. In Montreal, δO-18 would be about −15 per mil; on the Greenland icecap, it would average −30 per mil.

Two factors in addition to latitude make the isotopic composition vary at any given place on the surface of the globe. First, it varies seasonally. The isotopic composition of summer rains are heavier—higher in O-18—than winter rains. This is especially pronounced on the Greenland icecap, where the annual temperature cycle is very large. The isotopic composition also changes with altitude. In California, for instance, as air masses rise from sea level to the top of the Sierra Nevada Mountains, they cool and drop much of their moisture on the western

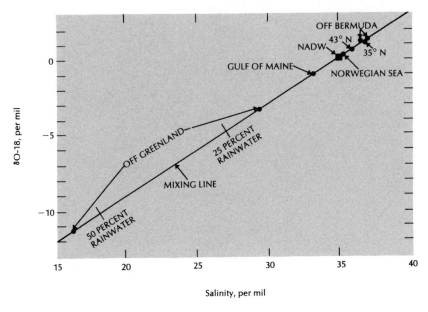

Figure 6-3 Plot of δO-18 versus salinity for samples from the North Atlantic Ocean. The differences can be largely accounted for by mixing various proportions of fresh water (salinity = 0 per mil) of isotopic composition −20 per mil with Gulf Stream water (salinity = 36 per mil) of isotopic composition +1 per mil. Water found in the deep Atlantic (NADW) has a composition falling on this line, suggesting that NADW consists dominantly of water sinking at the northern end of the Atlantic. (Data collected by Samuel Epstein, California Institute of Technology.)

side of the mountain range. The isotopic composition becomes progressively more deficient in O-18 at higher elevations.

The net result of this poleward decrease in the O-18 content of precipitation is that it tends to enrich the O-18 content of ocean water at low latitudes and to deplete the O-18 in water at high latitudes. Of course, surface currents carry water back and forth between the equator and the poles and thus tend to destroy the gradient in isotopic composition produced by precipitation. The net effect is therefore small, but a large enough range in composition exists (isotopic measurements can be made with extreme accuracy of ±.05 per mil) to allow surface sea water samples to be characterized by their O-18 content.

Figure 6-3 shows a δO-18 versus salinity diagram for sea water samples taken in the North Atlantic Ocean. Those samples with salinities of about 36 per mil (normal for that area of the ocean) have oxygen isotope compositions about 1 per mil higher in δO-18 than samples of SMOW. In the lower left-hand corner of the plot is a water sample taken close to Greenland. This sample is half fresh water and half sea water, so its salinity is only 16 per mil. It has an isotopic

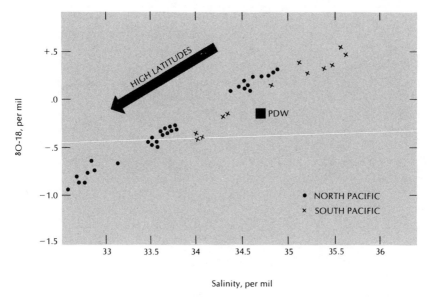

Figure 6-4 Plot of δO-18 versus salinity for surface water samples taken throughout the Pacific Ocean. Deep water in the Pacific cannot be generated from any mixture of these waters. (Data collected by Harmon Craig and Louis Gordon, both of Scripps Institution of Oceanography.)

composition of −11 per mil, or about a 12 per mil difference from the open ocean samples. The fact that all the North Atlantic Surface Water points fall along one line suggests that all these samples are mixtures of fresh water (which would have a salinity of zero and an isotopic composition of about −20 per mil) and water typical of the open ocean (salinity 36 per mil, δO-18 = 1 per mil). The influence of continental runoff dominates the relationship between salinity and oxygen isotope compositions in the high northern latitudes of the Atlantic.

North Atlantic Deep Water samples taken from 3000 meters deep all the way to the equator have salinities and oxygen isotope compositions that fall on the curve for North Atlantic Surface Water! Since NADW differs in composition from deep Pacific and Antarctic waters which do not fall along the North Atlantic Surface Water curve, any large admixtures of these waters would pull the NADW point off this line. We can only conclude that NADW must consist almost entirely of surface water from the North Atlantic, for if more than about 20 percent Antarctic Bottom Water, Antarctic Intermediate Water, or Mediterranean water were mixed with it, NADW would not fall on the O-18 salinity line.

Figure 6-4 shows the relationship between O-18 and salinity for Pacific surface water samples. The value for Pacific Deep Water is also given. Note that the samples from high latitudes are all lower in salinity and O-18 content than samples from the equatorial region.

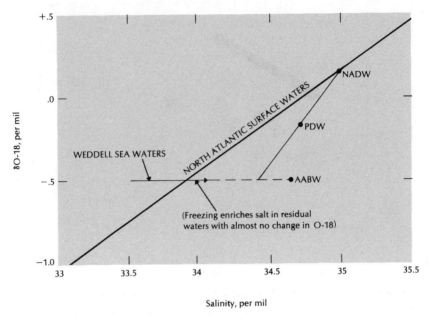

Figure 6-5   Relationship between the salinity and O-18 content of the major deep water types: NADW, AABW, and PDW. The water found in the deep Pacific can only be generated by mixing roughly equal parts of water originating in the northern Atlantic (NADW) and water originating along the edge of the Antarctic continent (AABW and related waters). (Data collected by Harmon Craig and Louis Gordon, both of Scripps Institution of Oceanography.)

This reflects two facts: the rains falling in high latitudes are depleted in O-18, and there is a net poleward transfer of water vapor through the atmosphere. The details of the pattern are, however, very complicated. The latitude effect is one important factor, but not the only one; variations related to local meteorological conditions and to current trajectories also exist.

The Pacific Deep Water point shown in Figure 6-4 does not fall on the North or South Pacific surface water lines, nor does it fall in between these lines. No single surface water sample in the Pacific Ocean or any combination of these waters could result in PDW. In fact, no single source of water in the entire surface ocean is identical in composition to PDW. Hence Pacific Deep Water cannot originate at one discrete point in the ocean. You can see from Figures 6-4 and 6-5 that PDW could be produced by mixing NADW with water sinking in the Antarctic (Antarctic Bottom Water or AABW). Actually, the deep water points of PDW, NADW, and AABW in Figure 6-5 are almost on a line.

Figure 6-5 also shows that waters collected from the Weddell Sea (along the coast of Antarctica opposite the South Atlantic) fall along the horizontal line on the O-18/salinity diagram. The reason this happens is that sea ice forms in this area during the winter. Since salt is not a component of ice, the salt remains behind in the residual water and the salinity of this water rises. However, when ice forms, almost no fractionation of the oxygen isotopes occurs. The O-18/O-16 ratio in the sea ice is nearly the same as the O-18/O-16 ratio in the water from which it formed. So the freezing process increases the salinity of the residual water without changing its O-18 content. This freezing process creates water that is more dense than any other water in the ocean. The water sinks down along the edge of the Antarctic continent into the deep Antarctic basin and flows out toward the Atlantic to form the Antarctic Bottom Water mass (the O-18/salinity point for this AABW mass falls along an extension of the Weddell Sea line, as shown in Figure 6-5).

Two surface sources account for the bulk of deep water in the ocean:

1. North Atlantic Deep Water falls along a line that defines the waters found at the surface of the North Atlantic
2. Antarctic Bottom Water falls along an extension of the Weddell Sea line

Although Pacific Deep Water does not *quite* fall on the line joining NADW and AABW (see Figure 6-5), other sources of water forming elsewhere along the Antarctic coast that are lower in salinity but that have the same O-18 as AABW can account for this discrepancy. Hence, each of the two known sources of deep water—the water sinking in the Norwegian Sea (NADW) and the water sinking along the Antarctic coast (AABW and its related waters)—have isotopic composition/salinity relationships consistent with that in the deep Pacific. A 50/50 mixture of these two would be identical to PDW.

So, one technique chemists have added to the physical oceanographer's tools for studying water masses is the use of the O-18 content of ocean water in conjunction with salinity as an indicator of the sources of subsurface water masses.

Another technique, which although not isotopic was discovered by geochemists in the course of isotopic studies, is a conservative tracer called NO. NO is calculated by adding 7 times the NO_3^- ion concentration to the O_2 gas concentration in a given water sample (of course, both concentrations must be in units of moles). Despite the fact that O_2 and NO_3^- concentrations change with time for parcels of water which descend from the surface into the body of the sea, the sum obtained from the above expression should not change. As we saw in Chapter 5, for every 100 moles of C burned during respiration roughly 20 moles of N were also burned (to produce NO_3^- ion). This requires the use of about 140 moles of O_2. Thus for each 7 moles of O_2

consumed 1 mole of NO_3^- appears. The sum $7\ NO_3 + O_2$ will not change as a result of respiration; the lowering in O_2 will be exactly compensated by the increase in $7\ NO_3^-$!

As shown in Figure 6-6, the subsurface water sources in the ocean have quite different NO values. Mediterranean outflow water, for example, has a very low NO value because of its relatively high temperature; the high density of this water is the result of high salt content rather than low temperature. On the other hand, waters of Antarctic origin have very high NO values because they are the coldest and because they leave the surface with unusually high NO_3^- contents. These differences permit NO versus temperature diagrams to be used in much the same way that we have already seen salinity versus O-18 diagrams used.

Vertical Mixing Rate Indicators

The other isotopic tools to be discussed have to do with methods of determining the mixing rate *within a given water mass*. So far, we have considered only the rate of water exchange between warm and cold waters. We will now examine the nature of the mixing processes within the separate reservoirs. In turn, we will discuss near-bottom, mid-depth, thermocline, and near-surface waters. Different isotopic techniques are used at each level.

Vertical Mixing in Near-Bottom Waters

We will begin with a method for studying the mixing processes that occur very near the bottom of the ocean. These processes are of considerable interest because they determine the nature of the interactions between the water and the sediments, but they are very difficult to study because the near-bottom region is so remote and the mixing rates there are so slow.

Oceanographers have gained some concept of the magnitude of near-bottom horizontal currents by using current meters. However, no physical device has been found that can be lowered to the bottom of the ocean in order to measure the vertical component of mixing (which is several orders of magnitude slower than the horizontal rate).

Fortunately, the sediments at the bottom of the ocean emit a radio-isotope, radon (Rn-222), which can be used to determine both the nature and rate of mixing in the lower 100 or so meters of ocean water. Radon is a member of the same U-238 decay chain that produces the tracer Th-230 discussed earlier. As we already know, U-238 is quite soluble in sea water, but its first major daughter product, Th-230, is extremely insoluble. Hence the Th-230 is rapidly removed to the sedi-

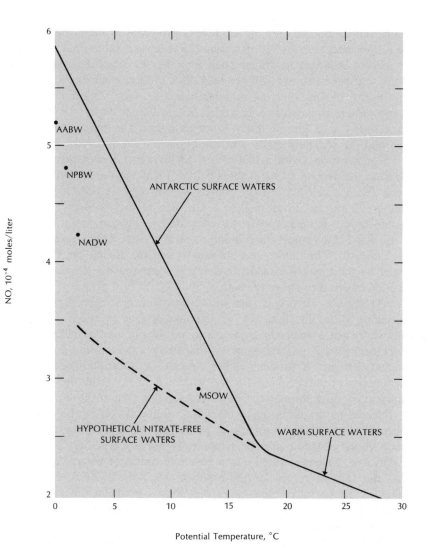

Figure 6-6 NO (O_2 + $7NO_3$) versus potential temperature for some of the key water masses in the ocean. AABW, NPBW, NADW, and MSOW refer respectively to Antarctic Bottom Water, North Pacific Bottom Water, North Atlantic Deep Water, and Mediterranean Sea Outflow Waters. Samples from the circumpolar surface waters of the Antarctic Ocean fall along the solid line rather than along the dashed extension of the line for nitrate-free waters. Only surface waters warmer than about 17° C prove to be nitrate-free. Colder waters contain measurable amounts of all the limiting nutrients. The complex factors controlling the amounts of this residual nitrate in very cold waters differ in different parts of the world. Hence deep waters from different sources have quite different NO values. Data collected as part of the GEOSECS study.

ments. An isotope of radium, Ra-226, is produced in the sediments by the further decay of Th-230. Radium-226 is fairly soluble; that is, about 10 percent manages to diffuse back into the sea from the pore fluids of the sediments. As we will see later, this radium itself has proved to be a useful tool in studying mid-depth mixing. The 90 percent or so of the radium that remains behind in the sediment decays to Rn-222. Since radon is a gas, it tends to migrate into the pore waters that surround the sediment grains and to diffuse into the overlying ocean.

Radiocarbon (with a half-life of 5700 years) moves through the water on the average of 8200 years before it is destroyed by radiodecay; radon (with a half-life of only four days) has only six days to wander. Radon-222 travels no more than a few tens of meters away from the bottom before it undergoes radioactive decay. This makes it possible to use Rn-222's vertical distribution in near-bottom waters as a mixing rate indicator. The nature and the rate of vertical mixing can be determined by the manner in which radon gas concentration drops off as a function of distance above the bottom.

At present, no one is sure what sort of oceanic mixing processes are taking place, but there are two hypotheses. One envisions organized cells where water descends in one place and rises in another (like boiling water). The other postulates eddies that carry water up and down in a random fashion. If random eddies operate, a diffusion model might characterize oceanic mixing. Before, when we talked about the diffusive transport of gases through the stagnant boundary film at the sea surface, we referred to the diffusion of individual molecules. It is also possible to speak of the diffusion of larger parcels of water traveling as units. Many grams of water may be temporarily moving together. This type of diffusive movement turns out to be many orders of magnitude faster than the rate of molecular diffusion. In fact, these diffusive mixing processes are so much more rapid that they overwhelm the motion of individual molecules to such a degree that the latter are insignificant except in the surface boundary film.

If the mechanism of mixing near the bottom is the diffusion of random eddies and if the effectiveness of this mixing does not change with distance above the bottom, then the amount of radon should decrease in a simple exponential manner away from the sediment–water interface, or:

$$[Rn]_x = [Rn]_0 e^{-ax}$$

where x is the distance above the interface, a is a constant determined by the eddy mixing rate, $[Rn]_0$ is the radon concentration in the bottommost water, and e is the base of the natural logarithms. The constant a is equal to the square root of the ratio of the decay constant of radon λ to the coefficient of eddy diffusivity k, or:

$$a = \sqrt{\frac{\lambda}{k}}$$

Actually, it is not quite this simple, for there are two components of radon in deep water. Some of the radon is produced by radium present in the sea water (the 10 percent which diffused from the sediments); the rest of the radon comes from the underlying sediment. If the Ra-226 content of the water is measured, the contribution of radon produced *in situ* can be computed and eliminated by subtraction. The radon remaining is the amount that has seeped out of the sediments. Its concentration should decrease exponentially toward zero with increasing distance from the bottom.

Under these circumstances the concentration of excess radon should halve at regular intervals above the bottom. In other words, beginning with the concentration immediately adjacent to the sediment, we could make radon measurements progressively upward in the water column until we found the point where the concentration of the radon moving up from the sediments was half the concentration at the interface. If this distance proved to be, say, 15 meters, then another 15 meters up (at 30 meters) the radon concentration ought to drop off by another factor of 2, to one-quarter of the original interface value; after another 15 meters (at 45 meters) the concentration would decrease by another factor of 2, to one-eighth the interface value; at 60 meters, it would decrease by another factor of 2, to one-sixteenth; and so on. The radon distribution could then be characterized by the vertical distance that must be traversed in order to observe a decrease of a factor of 2 in the Rn concentration from the sediments (see Figure 6-7). If radon is distributed in this manner, then the motions of near-bottom waters are being modeled successfully by a diffusive mixing process, and the rate of this mixing must be constant with distance above the bottom.

The distance $x_{1/2}$ over which the Rn concentration decreases by a factor of 2 can be related to the constant a as:

$$x_{1/2} = \frac{.693}{a}$$

Physically, a is the fraction of radon disappearing per unit of distance above the bottom. In other words, a has the same relationship to distance x that λ, the decay constant of a radioisotope, has to time. Since:

$$a = \sqrt{\frac{\lambda}{k}}$$

it follows that:

$$x_{1/2} = .693 \sqrt{\frac{k}{\lambda}}$$

Given λ for radon (2×10^{-6} parts per second) and x (from measurements of Rn-222 versus distance above bottom), the value of k (the coefficient of vertical eddy motion) can then be calculated as:

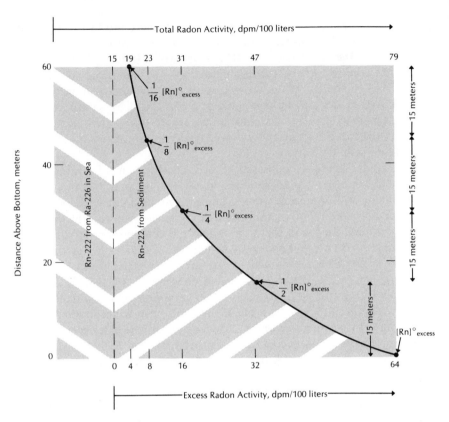

Figure 6-7   Ideal distribution of radon above the sea floor. The assumptions are made that mixing is by random eddies, that the mixing rate is independent of distance above the bottom, and that there is no influence of horizontal currents or topographic changes.

$$k = \left(\frac{x_{1/2}}{.693}\right)^2 \; \lambda$$

Thus, if the distance over which the Rn content halves ($x_{1/2}$) is 15 meters (1.5×10^3 cm), then k will be 9 cm^2/sec (6 orders of magnitude greater than the coefficient of molecular diffusion). If the halving distance were 7.5 meters instead, the eddy diffusion coefficient would be only one-quarter as great, or 2.2 cm^2/sec; if the distance were 30 meters, the coefficient would be four times as great, or 36 cm^2/sec.

Measurements made to date by this method often yield the vertical distribution predicted by the model (see the examples given in Figure 6-8). Values of k ranging from 2 to 200 cm^2/sec have been found in various parts of the ocean floor! Many of these measurements do not show the simple pattern of vertical decrease predicted by the random

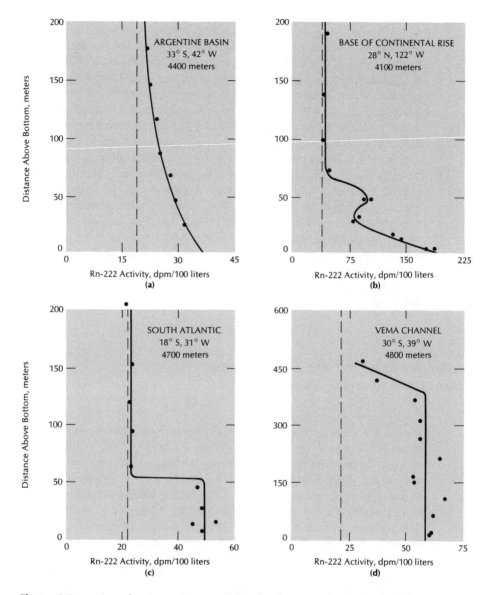

Figure 6-8   Actual measurements of the distribution of radon above the bottom. In each case, the contribution of radon from Ra-226 dissolved in deep sea water is indicated by a dashed line. Example (a) from the Argentine Basin shows the ideal distribution predicted if transport is by vertical eddies at a rate independent of distance above the sea floor. In example (b), taken from the base of the continental rise off northern Mexico in the Pacific Ocean, the secondary maximum in Rn concentration indicates that horizontal effects must be important. Example (c) from the South Atlantic shows that very rapid mixing occurs in the bottommost 55 meters; no measurable excess is found above this level. Example (d) from the Vema Channel, through which bottom water spills northward from the Argentine Basin, shows a layer of very rapid mixing extending 400 meters above the bottom. (The data presented here were collected by P. Biscaye (a), W. S. Broecker (b) and G. Mathieu ((c) and (d)), all of Lamont-Doherty Geological Observatory.)

eddy diffusion model, however. Instead, some show a nearly uniform upward distribution of radon from the sediment for the first 50–90 meters, followed by a very rapid decrease above this layer. This might be explained by a homogeneous layer in the near-bottom water (in which a rapid, organized overturn takes place) overlain by a stably stratified zone in which only eddy diffusion operates. Other measurements reveal a rapid decrease near the bottom, followed by a slower decrease at higher levels. This may reflect a situation where k increases with distance above the bottom. A few examples show a secondary maximum some distance above the bottom. Such a feature can only exist if horizontal transport is operative; water high in radon content must be moving in from the side. Much work remains to be done before these results can be properly interpreted, but this method does provide a quantitative way to study the vertical mixing rates near the bottom of the ocean. The results can be combined with data based on measurements of near-bottom horizontal flow to more precisely characterize the nature of the motions of the water near the bottom of the ocean.

Vertical Mixing in the Depth Range from 1000 to 4000 Meters

Figure 6-9 gives profiles of potential temperature (that is, *in situ* temperature corrected for adiabatic heating), salinity, and density as a function of depth in the Pacific Ocean. Although this profile was measured at 28° N latitude and 122° W longitude, it is fairly typical (except at less than 500 meters depth) of the entire Pacific. The first aspect of this profile we will consider is the salinity minimum at 1000 meters. The presence of such a layer requires the continuous injection of water of low salinity into this horizon. Otherwise, vertical mixing with both the overlying and underlying waters of higher salinity would gradually destroy the salinity minimum and produce a uniform gradient. The water which generates this minimum originates along the northern edge of the Antarctic Ocean. As it is not the most dense water produced in the ocean (both NADW and AABW are more dense), it sinks and spreads out at an intermediate density level.

The progressive temperature decrease and salinity increase from 1000 meters to the bottom require that a second source of deep water (the mixture of NADW and AABW we discussed earlier) is being continually injected along the bottom of the Pacific. Is the water from 1000 to 4000 meters in depth derived by mixing these two source waters? Our first clue can be found in Figure 6-10 which is a plot of

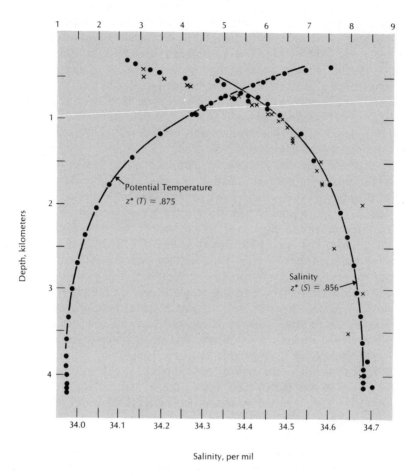

Potential Temperature, °C

Depth, kilometers

Potential Temperature
z* (T) = .875

Salinity
z* (S) = .856

Salinity, per mil

Figure 6-9 Plot of temperature (corrected for adiabatic heating), salinity, and density for a station located 28° N by 122° W about 250 miles southwest of San Diego in the Pacific Ocean. (Data collected by Harmon Craig, Scripps Institution of Oceanography.)

potential temperature versus salinity for the water samples taken in the depth region between the source levels. Note that the points on this graph fall along a straight line. Such a relationship is expected if this water is derived by mixing waters from the bounding source horizons. All mixtures of these two water types must have temperature–salinity (T-S) points falling along a straight line connecting the points designating the source waters. If, for example, there were a *third* important source adding water to the deep Pacific between 1000 and 4000 meters, then mixtures of this water with the other two would produce T-S points throughout a triangle defined by the T-S characteristics of the three water types. The straight-line relationship observed for this and for other Pacific profiles suggests that the major mass of deep water is comprised of a mixture of just two water types, intermediate (~1000 meters) and bottom (> 4000 meters) water. If a third source does exist, it must by chance have a T-S point that falls on the line connecting the T-S points of the intermediate and bottom waters. We will assume that no such coincidence has occurred.

The next question regards the nature of the processes which produce this mixing of the two source waters. Such mixing could be accomplished simply by the vertical movement of random eddies we discussed earlier, but this hypothesis must be rejected for two reasons. First of all, it provides no way to dispose of bottom water; this water is being continually added and must be removed *somewhere*. Second, if the rate of this random transfer of eddies were uniform over the 1000–4000 meter depth range, then not only would the combined T-S pattern be linear, but so would the actual changes in temperature and salinity with depth. Figure 6-9 reveals that this is clearly not the case: both properties show a concave shape over this depth region. Their rates of change with depth are most rapid at 1000 meters and become almost imperceptible at 4000 meters. Of course, this curvature could be the result of a change with depth in the rate of eddy mixing. If this mixing were far more rapid in the 3000–4000 meter range than in the 1000–2000 meter range, such a curvature would be generated. However, this would still leave the problem of disposing of the continually inflowing bottom water.

A better solution is to assume that a uniform upward flow of bottom water is superimposed on the random eddy mixing. Such a combination (upward advection* and diffusive mixing) would produce the observed concavity of the temperature and salinity profiles observed in Figure 6-9 and would also provide a means of disposing of inflowing bottom water.

If this is indeed the case, then the shape of the temperature or salinity profile is determined by the ratio of the coefficient of vertical eddy diffusion (in units of m^2/yr) to the upwelling velocity (in units of m/yr). For the deep Pacific, the shape can be shown to demand a value for this ratio of about 900 meters.

The problem is then to determine the absolute value of one of these rates. If this can be done, the value of the other can then be obtained from the restriction:

---

* Advection is comparable to the motion of a freight train; diffusion, to a drunken walk.

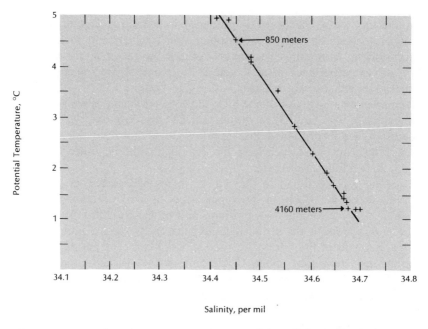

Figure 6-10   Plot of temperature (corrected for adiabatic heating) versus salinity for the station shown in Figure 6-9.

$$\frac{k\,(\mathrm{m^2/yr})}{w\,(\mathrm{m/yr})} \cong 900 \text{ meters}$$

where k is the coefficient of vertical eddy diffusion and w is the upwelling velocity.

We have already seen from the distribution of C-14 that the overall rate at which deep water is upwelling to the surface must be about 2 m/yr. If we assume that this figure also applies to the 1000–4000 meter range, we can then assign w a value of 2 m/yr. The value of k then has to be 4.5×10^2 m²/yr, or 1.5 cm²/sec.

A more direct approach is to consider the steady-state distribution of some radioisotope over the depth interval of interest. The isotopic concentration will be determined not only by the fraction of each water type present at any given level, but also by the extent to which the isotope has decreased in abundance due to radiodecay while trapped in this intervening zone. Plotted against salinity or temperature, the abundance of such an isotope would deviate from linearity. The deviation would be a measure of the time available for radiodecay and would therefore permit the absolute values of w and k to be disentangled.

Three radioisotopes appear suitable for this purpose: C-14 (half-life, 5700 years), Ra-226 (half-life, 1600 years), and Si-32 (half-life, 700 years). We have already learned how C-14 originates and how it enters the ocean. Silicon-32 is produced in a similar manner. Cosmic ray protons smash into the atmosphere and fragment the nuclei of the atmospheric atoms they encounter. Among the target atoms is Ar-40 (1 percent abundance in air), and Si-32 is one of the fragments produced. Like C-14, which reacts with O_2 to become CO_2 gas, the Si-32 atoms react with O_2 to become SiO_2 molecules (which, as we know, are not a gas). The $Si-32O_2$ molecules quickly associate with atmospheric aerosols, are incorporated into raindrops, and fall onto the land and sea. Those molecules reaching the sea mix with the other dissolved SiO_2 joining the marine cycle.

The Ra-226 found in the sea, as mentioned earlier, is produced by the decay of Th-230 in the upper layers of marine sediments. Since radium is more soluble than thorium, part of the Ra-226 is released to the pore waters in the sediment and manages to diffuse into the sea. Measurements of the ratio of Ra-226 to U-238 (the parent of Th-230) in the sea show that about 10 percent of the Ra-226 produced by the Th-230 in the sediment escapes. The rest remains trapped in the sediment until it disappears by radiodecay.

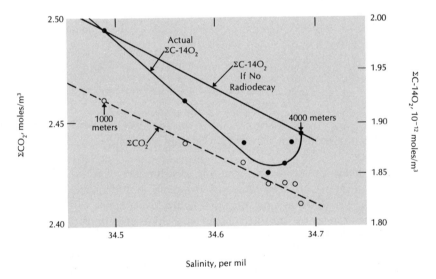

Figure 6-11 Plot of total dissolved inorganic carbon ΣCO_2 and total dissolved C-14 content $\Sigma C-14O_2$ versus salinity in the 1–4 kilometer depth range in the northeastern Pacific (28° N, 122° W). The straight line relationship for ΣCO_2 suggests that the influence of falling debris is unimportant. The minimum in $\Sigma C-14O_2$ is the result of radiodecay. If the rate of mixing were very rapid compared to the rate of radioisotope decay, then the C-14 points should fall along the indicated straight line. (The ΣCO_2 data collected by Raymond Weiss, Scripps Institution of Oceanography, and the C-14 data collected by Göte Ostlund, University of Miami.)

The applications of all three of these tracer isotopes share a common complication. They are removed from surface water by organisms, fall to the deep sea with particulate debris, and are returned to solution. Some of the destruction of this falling debris occurs within the 1000–4000 meter interval. This input tends to counterbalance the decrease in concentration produced by radioactive decay. If we are to use these tracers to establish absolute rates of vertical advection and eddy diffusion, we must take into account the input produced by the destruction of falling particles.

This can be accomplished for C-14 by considering the profile of normal (nonradioactive) dissolved carbon. The deviation of the normal carbon distribution from the distribution expected from the mixing processes themselves must be due solely to the *in situ* release of carbon from particles. Since we know the C-14/C ratio in surface water (and hence in the particulate debris falling from the surface), we can use the stable carbon excess at any given depth to determine the C-14 added to that level in the water column by the destruction of falling particles. Figure 6-11 shows the ΣCO_2 distribution measured in the 1000–4000 meter range in the eastern North Pacific plotted against salinity. No curvature is present, indicating that the oxidation of organics and the solution of $CaCO_3$ do not add significant ΣCO_2 (or $\Sigma C\text{-}14O_2$). Hence the observed concavity in the $\Sigma C\text{-}14O_2$ distribution (see Figure 6-11) must reliably reflect the loss of C-14 by radioactive decay. Figure 6-12 shows the distribution of C-14 expected for various values of the upwelling velocity w. Experimental data are shown for comparison. An upwelling velocity of 5 meters per year and an eddy diffusion coefficient of 4 cm²/sec provide the best fit to the experimental data.

An exactly analogous procedure could be followed for Si-32. By comparing the stable Si profile with the profile expected from mixing alone, we can estimate the rate of Si input from the *in situ* solution of diatoms and radiolarians. From a knowledge of the Si-32/Si ratio in near-surface water (and hence in the opal formed by these organisms), we can also estimate the amount of Si-32 added throughout the 1000–4000 meter depth range. Unfortunately, this method has not yet been exploited. The problem is that the amount of radioactive Si-32 in sea water is exceedingly small. In order to obtain enough to permit an adequate measurement, 100,000 liters of water must be processed! Bringing this much water from its depth to the deck of a ship proved impractical, and a device has been developed to extract Si from the water at depth. Within the next few years, this method will provide a means of cross-checking the results obtained by C-14 tracing. Because of its shorter half-life, Si-32 versus depth plots will show a greater "bulge" than C-14 versus depth plots and should therefore provide a more accurate means of obtaining the upwelling velocity.

There is no stable isotope of the element radium. Fortunately,

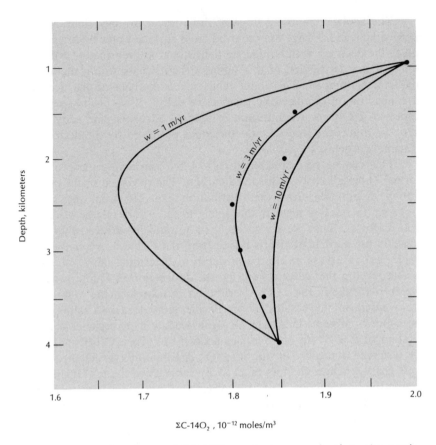

Figure 6-12   Plot observed $\Sigma C\text{-}14O_2$ content versus depth in the north-eastern Pacific Ocean (28° N, 122° W). Given for comparison are curves showing the expected distribution for upwelling velocities of 10 m/yr, 3 m/yr, and 1 m/yr. In each case, the influence of the *in situ* oxidation of falling organic debris and of the solution of falling $CaCO_3$ is assumed to be negligible (see Figure 6-11).

however, radium and barium have nearly identical chemical properties. Organisms appear to treat Ra-226 as if it were an isotope of Ba, as shown by vertical profiles of these two elements in the Pacific Ocean (Figure 6-13). Both Ra-226 and Ba show a roughly fourfold lower concentration in surface water than in deep water. The pattern of increase with depth is also nearly identical! The few measurements that have been made on particulate matter sieved from the sea show Ra-226 and Ba to be in nearly the same proportions as they are in surface water.

If this chemical similarity proves universal, then the distribution of Ba with depth in the sea will provide a means of correcting the Ra-226 distribution for the *in situ* addition through the destruction of particles

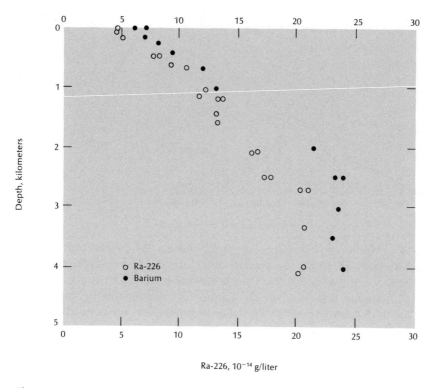

Figure 6-13 Depth distribution of Ra-226 and Ba in the northeastern Pacific Ocean (28° N, 122° W). Note that for both elements surface water shows a fourfold lower concentration than deep water. Both also show a more or less linear decrease down to 2500 meters. Below 2500 meters, Ra-226 and Ba have nearly constant concentrations. (Data for Ba collected by K. Wolgemuth, and data for Ra-226 collected by W. S. Broecker, both of Lamont-Doherty Geological Observatory.)

bearing Ra-226. Results for one profile treated in this way yield an advection velocity of 4 m/yr (a value consistent with that of 5 m/yr based on the C-14 and ΣCO_2 data for the same profile).

Vertical Mixing in the Main Thermocline

The main thermocline is the region of the ocean that has the greatest density gradient. Vertical mixing rates in this depth range might therefore be expected to be the smallest. The progress made to date in study-

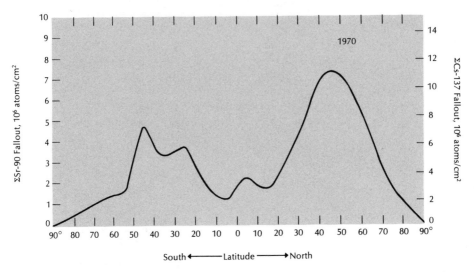

Figure 6-14   Latitude dependence of Sr-90 and Cs-137 fallout from nuclear weapons testing as of 1970. The main features are the higher average for the northern hemisphere (because most of the tests were made there) and the maxima at about 45° latitude (the result of the transfer of stratospheric air through the "gap" in the tropopause located in this latitude range). The total amount of Sr-90 reaching the surface at any given latitude has been estimated by sampling soils.

ing this problem involves the use of man-made radioactive "dyes" Cs-137, Sr-90, C-14, and H-3. The term "dye" is used because in this case the radioisotopes are used as dye tracers rather than as radioactive time clocks. Their radioactivity merely allows these isotopes to be detected in very small quantities. The procedure is analogous to sprinkling a red dye over the surface of the ocean and observing its distribution with time. Mankind has produced great quantities of these four tracers by testing nuclear weapons in the atmosphere. Because the half-lives of the isotopes are quite long, they will be with us for some time to come.

The four isotopes Cs-137, Sr-90, C-14, and H-3 were produced mainly by large thermonuclear tests and were carried into the stratosphere along with the clouds of hot gas generated by these explosions. The stratosphere, as its name indicates, is thermally stratified: it becomes warmer (and hence less dense) with increasing elevation, just like the oceanic thermocline. Thus there is not much tendency for air in the stratosphere to undergo vertical mixing. Because of this inherent stability, it took several years for all the radioactivity added to the stratosphere to migrate down into the underlying troposphere. The troposphere is, by definition, a zone of vertical instability (*tropo =* turning). Mixing from the top to the bottom of this lower region of the

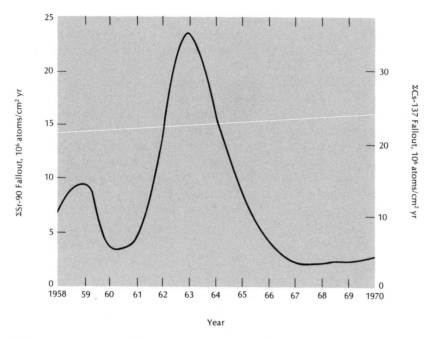

Figure 6-15 Annual rate of Sr-90 and Cs-137 fallout onto the earth's surface as a function of time. The rapid decline after 1963 reflects the moratorium in nuclear testing. The rates given are global averages.

atmosphere occurs in a matter of a few days. Once radioactive dust enters the troposphere, it is readily scavenged by raindrops and carried down onto the earth's surface. Knowing that sprinkling radioactivity over the world was a dangerous business, the Atomic Energy Commission carefully monitored the extent of the radioactivity by measuring air, soil, and rain samples taken at various places on the continents and on oceanic islands. Thus we have quite a concise record of the input function of these isotopes to the earth's surface (see Figures 6-14 and 6-15). This material from the stratosphere was deposited on the surface of the ocean over a period of about a decade (1954–1964) and has since been mixed downward by the normal processes operating within the sea. It is as though we had, over a period of ten years, sprinkled dye in a known amount over the surface of the ocean.

Of the four radioactive dye tracers, only C-14 is not incorporated into rain. Instead, as we have seen, C-14 enters the ocean via gas exchange. Once in the sea, its vertical distribution, like that of the other three isotopes, is determined by normal oceanic mixing processes. Fortunately, distribution of the radiocarbon produced by cosmic rays was already reasonably well determined by the time the bomb-created C-14 began to enter the ocean, so we can subtract the natural contribution from the total and simply use the resulting bomb-produced portion as the tracer. (The naturally occurring tritium concentration in the ocean is so low that it can be neglected with respect to bomb-produced H-3; neither Sr-90 nor Cs-137 was present in the ocean prior to the manufacture of nuclear devices.)

Now let us consider the results we have obtained thus far. The concentrations of H-3 and C-14 in the equatorial region of the Atlantic Ocean are shown in Figure 6-16. Ten years after the dispersal of these isotopes over the ocean began, we find that radioactivity in the equatorial area is confined to the upper 60 or so meters! The top of the thermocline in the equatorial regions is very shallow and the density gradient within this thermocline is very great. Complete mixing of fall-out occurs to only a few tens of meters, and no measurable amounts of radioactive isotopes are found below this depth. Further north in

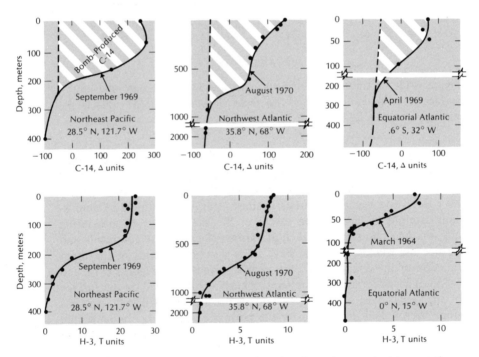

Figure 6-16   Depth profiles for bomb-produced radiocarbon and tritium at three places in the ocean. In the case of C-14, the bomb component is obtained by subtracting the pre-bomb natural C-14 profile from the measured profile. The units used for the C-14 measurements are per mil difference from a standard. For tritium, the unit is one tritium atom per 10^{18} hydrogen atoms. Clearly the depth of penetration shows large geographic variations. (These measurements were made by Göte Ostlund, University of Miami, and Karl Otto Münnich and Wolfgang Roether, both of Heidelberg University.)

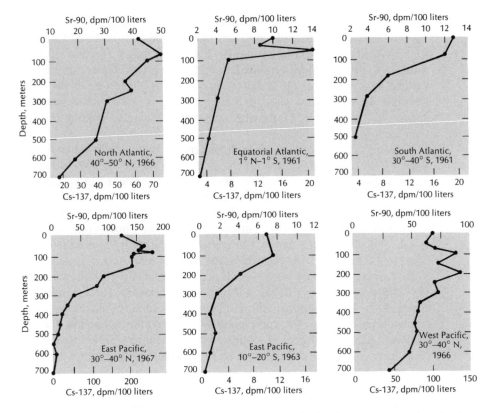

Figure 6-17 Distribution of Cs-137 and Sr-90 with depth at various points in the ocean. The depths to which these isotopes have penetrated show large geographic variations. (Data obtained from *Radioactivity in the Marine Environment,* published by the National Academy of Science, 1971.)

the Atlantic Ocean, we encounter ever colder winter waters with increasingly greater tendencies for vertical mixing. The thermocline becomes deeper and the density gradient within it becomes much smaller. The fallout isotopes spread over a greater depth range. In the latitude range of 35°–45° N and at a 700 meter depth, the radioactive concentration is about half as great as it is at the surface (again, see Figure 6-16). Similar distributions are shown for Cs-137 and Sr-90 in Figure 6-17.

One of the most exciting discoveries resulting from these "dye" studies was made in the North Pacific: there, both H-3 and Cs-137 showed a marked maximum at a depth of about 100 meters by workers at Scripps Institution of Oceanography. This maximum corresponds to a well-known salinity minimum in the near-surface Pacific Ocean which is produced by the sinking and lateral spreading of waters from the extreme northwest corner of the Pacific Ocean, north of Japan. Since the radioactivity maximum was observed as far away from this source as 1000 miles, we infer that in ten years water tagged with H-3 and Cs-137 isotopes has moved along its density horizon a major fraction of the width of the Pacific Ocean basin! During this long trip, remarkably little vertical mixing occurs. (If it did the maximum could not be detected.) Laminae of water that slide out along horizons of constant density appear to be capable of moving enormous distances without losing their identity.

Using the distribution of bomb-generated radioactivity as a function of both time and location in the ocean, we should be able to learn much about the nature of these horizontal spreading phenomena. It appears that water within the thermocline region is renewed by horizontal spreading. Blobs of water a little more dense than the underlying water form somewhere on the surface, sink to their appropriate density levels, and spread out across the ocean. That this is a general phenomenon is revealed by records made with a new device called the STD. Lowered into the sea, this device continuously records salinity (S) and temperature (T) as a function of pressure (and hence of depth, D). These records show many little salinity maxima (and minima) matched with corresponding temperature maxima (and minima), depicting waters of slightly different temperature and salinity (but nearly the same density), which at various times sank and spread without loss of identity.

The combined use of the radioactive "dye" tracers produced by nuclear testing and of continuous temperature and salinity profiles should elucidate the processes that are taking place in the main thermocline region of the ocean, where the strongest barrier to vertical mixing exists. Details of these processes have, until now, remained well beyond the oceanographer's grasp. Radioactive tracer studies will provide information about mixing processes operating in this region that cannot be obtained by studies of density distribution and horizontal currents alone.

Vertical Mixing in Near-Surface Waters

Measurements of the salinity and temperature of waters near the ocean surface always reveal a layer in which these properties are completely uniform. Depending on the geographic location and the season, the thickness of this layer ranges from a few tens to a few hundreds of meters. Its very existence requires that vertical mixing rates far more rapid than those found at greater depths must prevail. Are there chemical tracers which can be used to determine exactly how this rapid mixing takes place?

So far, only one shows promise. Radon-222, the same isotope which served as a tracer in near-bottom waters, can also be used to trace mixing rates near the surface. In near-surface waters, however, instead of an excess of radon over that expected from the radiodecay of the Ra-226 present, a deficiency is found. This is because some of the radon generated by the dissolved Ra-226 in the water escapes to the atmosphere. As the atmosphere over the ocean has a very low Rn content, the escaping radon is not replaced by a corresponding influx from the air (as is true for the other gases we have discussed). The magnitude of this Rn deficiency decreases with depth until eventually it reaches zero. At a great enough distance from the interface, the Rn content must just match the radon produced by the radium dissolved in the water. The manner in which the Rn deficiency changes with distance from the air–water interface depends on the rate of vertical mixing. Thus, in a manner analogous to that used in near-bottom waters, the distribution of radon with depth in the so-called "mixed" layer can be used to determine the coefficient of vertical eddy diffusivity. Because the Rn deficiencies are small ($\sim 40 \pm 20$ percent at the surface), this method cannot be exploited satisfactorily until measurements can be made within an accuracy of 5 percent or better, and techniques to achieve this have only recently been developed. (One example of the application of this method is shown in Figure 6-18, page 172.)

Horizontal Mixing Rate Indicators

Because of their relatively uniform delivery to various parts of the ocean surface, isotopes produced by cosmic rays and by nuclear testing are not useful indicators of *horizontal* mixing rates in the surface ocean. An isotope produced by the decay of Th-232 is suitable for this purpose, however. Radium-228 (half-life, ~ 7 years) is being released to the surface ocean from the shallow water sediments along the continental margins where sizable quantities of thorium, its parent, reside. (The situation is analogous to the release of Ra-226 produced by the Th-230 in sediments.) Horizontal mixing processes carry Ra-228 toward the interior of the ocean basin. The main current systems move the water around in great hemispheric gyres. In the North Atlantic, the Gulf Stream is the western limb of such a system. It is clear that these currents carry material around the basin, but it is not clear how rapidly the mixing proceeds radially. Radial mixing transports pollutants from the coasts to the interior of the ocean basins. As this mixing is through turbulent eddies produced by the main current, it is possible to characterize radial mixing by the diffusion of random eddies. The distribution of Ra-228 enables us to determine the magnitude of the coefficient characterizing this process. The gradient of Ra-228 away from the continent is a measure of this coefficient. A rapid decline away from the continental margin would indicate a rather slow mixing rate; a slow decline, a rather high mixing rate.

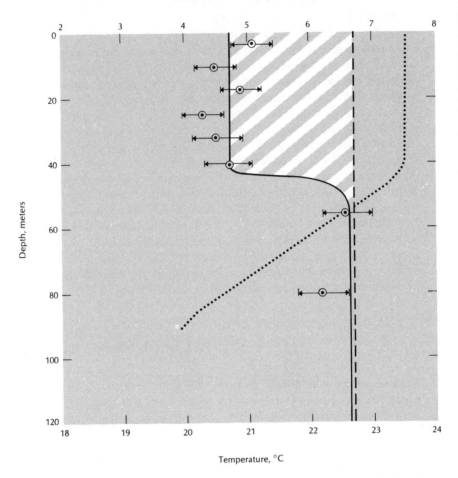

Figure 6-18 Concentration of radon gas as a function of depth at Atlantic GEOSECS station 57 (24° S 35° W). The concentration expected if no radon were escaping to the atmosphere is also shown as a dashed line. The difference between this equilibrium curve and the observed curve is a measure of the amount of radon lost to the atmosphere. The shaded area is the total amount of radon lost to the air. From the amount of missing radon and its depth distribution, it is possible to determine the piston velocity for gas exchange and to place limits on the rate of vertical mixing in the upper few tens of meters of sea water. The dotted line is the temperature profile at this station. (Data obtained by Guy Mathieu, Lamont-Doherty Geological Observatory.)

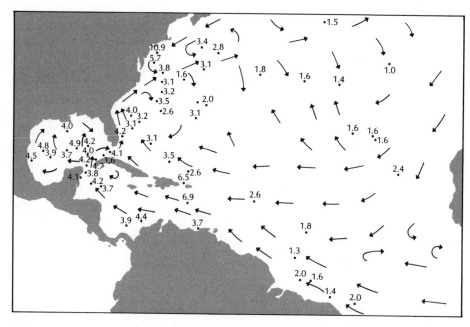

Figure 6-19   Map of the North Atlantic Ocean showing the distribution of Ra-228 in surface water. The concentrations are given in dpm/1000 liters. As Ra-228 (half-life ~7 years) is being released from sediments on the continental shelves, the rate at which its concentration decreases away from the coast is a measure of the rate of horizontal mixing. (Data obtained by A. Kaufman, R. Trier, and H. Feely, all of Lamont-Doherty Geological Observatory.)

Figure 6-19 is a map of the available data for Ra-228 concentration in the central North Atlantic Ocean. Figure 6-20 is a plot of Ra-228 activity versus distance from the continental margin along a traverse southeast from Long Island. The solid lines in Figure 6-20 are the distributions predicted for various constant rates of transport by horizontal eddy diffusion toward the interior of the ocean basin. An eddy diffusion coefficient for horizontal mixing on the order of 1×10^6 cm^2/sec is required to explain the observed distribution of Ra-228. This is 11 orders of magnitude greater than the coefficient for molecular diffusion and 6 orders of magnitude higher than the coefficient for vertical eddy diffusivity in the 1000–4000 meter depth range.

The combination of bomb-produced H-3 (tritium) and the stable product of its radioactive decay He-3 form a potentially powerful tool for measuring the rates of horizontal flow in the deep sea. A research team headed by Byron Clarke at McMaster University in Canada developed the capability to measure the very small amounts of He-3 generated within the sea by the decay of tritium. For example, tritium-tagged water which left the surface 12 years ago will, as of today, have

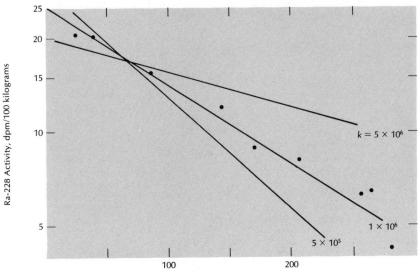

Figure 6-20   Plot of Ra-228 activity in North Atlantic surface water as a function of distance from the edge of the U.S. continental margin. The solid lines show the distributions expected for eddy diffusion coefficients of 5×10^5, 1×10^6, and 5×10^6 cm²/sec. (Data obtained by A. Kaufman, R. Trier, and H. Feely, all of Lamont-Doherty Geological Observatory.)

lost half its tritium atoms through radioactive decay. These atoms have become He-3. Thus the ratio of excess He-3 (over the atmospheric amount and ridge crest "smoke," if any) to H-3 would be unity in this water. This He-3/H-3 ratio will be less in "younger" waters and more in "older" but post-nuclear testing waters. As mentioned previously, tritium-tagged waters are currently sinking in the northern regions of the Atlantic and spreading to the south. This has been going on since 1954. The time elapsed since these post-nuclear waters left the surface is being neatly recorded by the He-3/H-3 ratio. Mixing between these waters and pre-nuclear waters (free of both H-3 and its He-3 product) will alter neither this ratio nor the He-3/H-3 age.

Summary

Isotopic tracers offer a means of studying mixing processes taking place within the ocean. The distribution of the stable trace isotope O-18 within the sea provides valuable evidence regarding the origin of waters in the deep sea. Natural radioisotopes produced by cosmic rays (C-14 and

Si-32) and by the decay of uranium and thorium (Ra-226, Ra-228, and Rn-222) can be used to determine the absolute rates of diffusive and advective mixing within the sea. Man-made radioactive isotopes (C-14, H-3, Sr-90, and Cs-137) provide powerful tracers for the mixing processes taking place within the main thermocline. Although the potential of these isotopes is clear and the techniques for their measurement are available, oceanwide surveys of their abundances are only now underway. When these surveys are complete, we will be able to meaningfully interpret what radioisotopic distributions have to say about the mixing processes within the ocean.

SUPPLEMENTARY READINGS

■ Articles dealing with the use of isotopes as water mass indicators:

Epstein, S., and Mayeda, T. "Variation of O-18 Content of Waters from Natural Sources." *Geochim. Cosmochim. Acta* 4 (1953):213–24. The original discovery of this powerful technique.

Craig, H., and Gordon, L. I. "Isotopic Oceanography: Deuterium and Oxygen-18 Variations in the Ocean and the Marine Atmosphere." *Proc. Symp. Mar. Geochem., Univ. Rhode Island Occ. Publ. 3-1965* (1965): 277–374. A discussion of the origin of AABW and of the factors influencing the distribution of O-18 within the sea.

■ Articles dealing with the use of radon as an indicator of near-bottom mixing:

Broecker, Wallace S., Cromwell, J., and Li, Yuan-Hui. "Rates of Vertical Eddy Diffusion Near the Ocean Floor Based on Measurements of the Distribution of Excess Rn-222." *Earth Planet. Sci. Letters* 5 (1968): 101–105.

Chung, Yu-Chia, and Craig, H. "Excess Radon and Temperature Profiles from the Eastern Equatorial Pacific." *Earth Planet. Sci. Letters* 14 (1972): 55–64.

Both articles discuss the measurement of excess radon in near-bottom waters and its implications with regard to near-bottom mixing rates.

■ Articles concerning vertical advection and eddy diffusion rates in the deep ocean:

Craig, H. "Abyssal Carbon and Radiocarbon in the Pacific." *J. Geophys. Res.* 74 (1969):5491. Gives the mathematical basis for calculating rates of vertical mixing from the distributions of a radioisotope and its stable chemical equivalent.

Somayajulu, B. L. K., Lal, D., and Craig, H. "Silicon-32 Profiles in the South Pacific." *Earth Planet. Sci. Letters* 18 (1973):181–88. Discusses the distribution of Si-32 in the deep Pacific and its potential as an indicator of vertical mixing rates in the deep sea.

Chung, Yu-Chia, and Craig, H. "Radium-226 in the Eastern Equatorial Pacific." *Earth Planet. Sci. Letters* 17 (1973):306–18. Discusses the distribution of Ra-226 in the deep sea and the potential of the Ra-226/Ba pair as indicators of the rates of vertical mixing in the deep sea.

■ Publications dealing with radioisotopes produced during nuclear weapons testing and their potential as tracers of thermocline mixing processes:

Radioactivity in the Marine Environment. Committee on Oceanography. (Washington, D.C.: National Academy of Sciences, 1971). A book summarizing our knowledge of radioactivity in the marine environment.

Ostlund, Göte, Rinkel, M. O., and Rooth, C. "Tritium in the Equatorial Atlantic Current System." *J. Geophys. Res.* 74 (1969):4535–43. Discusses the mixing rates in the main thermocline in the North Atlantic Ocean as determined from the distribution of tritium.

Silker, W. B. "Beryllium-7 and Fission Products in the GEOSECS II Water Column and Applications of Their Oceanic Distributions." *Earth Planet. Sci. Letters* 16 (1972):131–37. Discusses the use of cosmic ray produced Be-7 as an indicator of the rates of vertical mixing in the upper regions of the main oceanic thermocline.

PROBLEMS

6-1 The following results were obtained for a profile of radon measurements in near-bottom waters:

Distance above Bottom, meters	Radon Activity, dpm/100 liters
200	20.3
100	19.8
50	20.7
40	21.2
30	24.1
20	36.3
10	84.0

What concentration of radon is produced by the Ra-226 dissolved in the water column? (Assume that the concentration is uniform over the sampling interval.) What is the apparent coefficient of vertical eddy diffusivity as determined from the distribution of excess (sediment-derived) radon?

6-2 A subsurface water has a δO-18 value of − .3 per mil and a salinity of 34.1 per mil. If the water is 50 percent NADW by volume (δO-18 = .15 per mil, S = 34.95) determine the salinity and the δO-18 for the component that comprises the other 50 percent.

6-3   What is the value of the vertical eddy diffusion coefficient in a water column which has an upwelling rate of 10 m/yr and a z^* value of 350 meters? (Assume that the one-dimensional advection–diffusion model is applicable.)

6-4   The following measurements were obtained for a vertical profile in the Pacific Ocean:

Depth, meters	Salinity, per mil	Potential Temperature, $°C$	O_2	NO_3	SiO_2
			←	10^{-6} moles/liter	→
1060	34.30	3.33	28	36.6	115
1400	34.44	2.55	23	37.4	155
1800	34.54	2.07	38	40.3	169
2300	34.61	1.67	76	39.3	173
2880	34.65	1.38	113	38.1	170
3130	34.66	1.29	126	37.4	165
3380	34.68	1.23	135	36.5	162
3630	34.67	1.18	142	35.9	159
3880	34.68	1.15	147	35.8	156
4170	34.68	1.12	151	35.4	158
4470	34.68	1.11	152	35.3	159

Do these measurements provide any evidence for *in situ* respiration or opal dissolution? If so, state the evidence.

6-5   If a near-bottom water sample from the North Atlantic collected in 1976 yields an excess He-3 to H-3 ratio of 3.0, when did this water leave the surface? Is this result consistent with what we know about the He-3/H-3 method?

6-6   A pair of surface and thermocline samples, selected because they have the same salinity and temperature, yield the following nutrient data:

	Potential Temperature, $°C$	Salinity, per mil	O_2	NO_3	NO
				10^{-6} moles/liter	
Surface	13.0	35.30	259	1.0	266
Subsurface	13.0	35.30	115	21.5	266

What conclusion can you draw from these results?

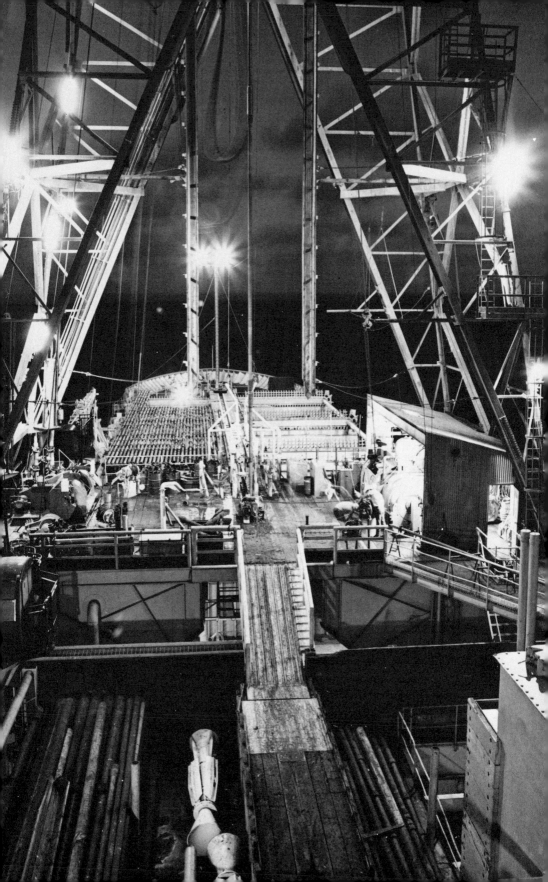

7

TEMPORAL CHANGES, PAST AND FUTURE

Thus far, our main consideration has been how the ocean operates today. Only a few brief comments have been made about the ocean's past and nothing has been said about its future. What changes has the chemistry of sea salt undergone in the past? Did these changes have an important influence on the course of life? Are the pollutants we are so rapidly pouring into the sea changing its chemistry? If so, what influences will these changes have on marine organisms and on man? To answer these questions we must add the dimension of time to our discussion of marine chemistry.

Many scientists suspect that when the earth was formed it had little or no atmosphere or ocean, that the water and other volatile substances now found on the earth's surface were distributed more or less uniformly throughout its interior. If the water currently in the ocean were mixed with the entire substance of the earth, by weight it would constitute about one-quarter of 1 percent of the total mixture. Since some water may still remain in the earth's interior, we can conclude that the materials that accumulated to form the earth must have had an average water content by weight of at least 250 ppm. Indeed, meteorites (fragments of small planetary objects) that fall onto the earth today do contain about this amount of water.

But when did the water built into the earth migrate to the surface? One popular but unproven hypothesis is that the earth's continental crust, its atmosphere, and its oceans evolved together over the course of geologic time—that the volcanic material issued forth from the earth's interior added to the amount of continental material, to the amount of atmospheric gas, and to the amount of ocean water. Energy released by the radioactive decay of uranium, thorium, and potassium heats the earth's interior and drives this process. It is thought that the earth's upper mantle has remained at a temperature close to its melting point. In fact, some regions of the upper mantle have been (and still are) partially liquified. The important point is that the first fraction of interior material to liquify does not have the same chemical composition as the whole. Volatile materials such as water and nitrogen are preferentially included in the "juice." So, too, are the large ions that do not fit well into the lattices of earth-interior minerals (iron and magnesium silicates). Thus at the surface of the earth an enrichment occurs, not only in substances like nitrogen (as atmospheric N_2) and hydrogen (as oceanic H_2O), but also in potassium, rubidium, barium, cesium, uranium, the rare earths, and many other "bulky" elements.

If this hypothesis is correct, then dating the growth of the continents will also give us some idea of when the ocean and the atmosphere were formed! Radioactive age determinations reveal that the oldest rocks on the surface of the earth are about 3.8 billion years old (compared to an age of 4.5 billion years for the earth itself). We have no record of the first .7 billion years of earth history. Perhaps it took this long for the radioactive elements contained by the earth to heat the interior to the point where melting occurred! Between 2.8 billion and 2.5 billion years ago, a tremendous amount of the earth's present continental crust was formed; much of Canada, Australia, Scandinavia, and Brazil—in fact, 25 percent of all continental material—formed during this period. Since then, several other episodes of continental growth have supplied the remaining material, and it is likely that the ocean increased in size during each episode of continental growth.

Attempts have been made to see if any volatiles are still being discharged from the interior. Water from hot springs and steam from volcanic vents are good prospects. Studies have shown, however, that both consist mainly of water from the earth's surface rather than from its interior. Rain water seeps down into the crust and reacts with the sediments through which it percolates. If, in its course, it encounters a pocket of hot rock, the water is heated and driven back to the surface through a volcanic vent or the orifice of a hot spring. If a component of *interior* water is being discharged with this recycled water, no one has yet been able to identify it. It constitutes no more than a few percent of the total.

As we pointed out in Chapter 5, recent measurements of rare gases in sea water have shown that the helium in deep sea water must, in part, be derived from the earth's interior. More helium is dissolved in sea water than can be accounted for by the amount of helium in the atmosphere. As He-4 is continually generated within the earth from the alpha particles given off during the decay of uranium and thorium (and their daughter products), we cannot safely use this excess as an index of the outgassing of original volatiles. For even if the earth were thoroughly outgassed of all the primary volatile constituents, He-4 would still be discharged from the interior because of its continual renewal. Helium-3, however, is not regenerated by radioactive decay within the earth. Since an even greater excess of He-3 (with respect to the amount expected from the atmosphere) is present in the deep ocean, it appears that a primary constituent of earth matter is still leaking to the surface after 4.5 billion years of entrapment in the interior! Possibly the escape rate of He-3 will eventually allow us to compare the current rate of water loss from the interior with that required to produce the ocean over geologic time.

Factors Controlling the Anion Content of the Sea

How did the salt content of ocean water evolve? First, let us look again at the major anions. We have seen that chloride is the most abundant anion, that sulfate is the second most abundant, and that bicarbonate runs a very poor third. As these three substances do not find happy homes in igneous rock, when liquids from the interior reach the surface and crystallize, any chlorine, sulfur, and carbon they contain are largely released either in volatile form to the atmosphere or in dissolved form to the surface waters. So the place of residence on the earth's surface for these three elements is either in the sea itself or in chemical precipitates formed within the sea and buried in the sediments; only a bit of the carbon and no chlorine or sulfur are present in the atmosphere. Inventories of the amounts of these chemical precipitates show that roughly half of the chloride currently resides in sodium chloride (NaCl) deposits in sediments, that about three-quarters of the sulfur is present in sulfate and sulfide minerals in sediments, and that only about .2 percent of all of the carbon released from the earth's mantle is now dissolved in the sea (the remaining carbon, about 500 times more, resides in sediments as limestone ($CaCO_3$), dolomite ($MgCa(CO_3)_2$), and organic residues (see Table 7-1).

If we assume that as water emerged from the interior it carried chlorine, sulfur, and carbon in roughly constant amounts, then by the time half the water had reached the earth's surface, half the chloride, half the sulfur, and half the carbon had also arrived. If this is correct, then the salinity of the ocean need not have been drastically different at any time in the past.

Have the proportions of these elements contained in the oceans compared with the proportions contained in the sediments changed

Table 7-1 Distribution of the volatile elements among the atmospheric, oceanic, and sedimentary reservoirs. Amounts are given in moles of the element contained by the reservoir.

Element	Hydrogen	Carbon	Chlorine	Nitrogen	Sulfur
ATMOSPHERE					
Form	H_2O	CO_2	—	N_2	—
Amount	4×10^{18}	5×10^{16}	—	3×10^{20}	—
OCEAN					
Form	H_2O	HCO_3^- $+ CO_3^{--}$	Cl^-	NO_3^-	SO_4^{--}
Amount	8×10^{22}	3×10^{18}	7×10^{20}	4×10^{16}	4×10^{19}
SEDIMENT					
Form	H_2O	$CaCO_3$ $+ MgCa(CO_3)_2$ Kerogen	$NaCl$	Kerogen	$CaSO_4$ FeS_2
Amount	4×10^{21}	3×10^{21}	8×10^{20}	2×10^{15}	2×10^{20}
TOTAL					
Amount	800×10^{20}	30×10^{20}	15×10^{20}	3×10^{20}	2×10^{20}

with time? At present, the ocean is far from saturated with respect to any salt capable of precipitating chloride (Cl^-) ion or sulfate (SO_4^{--}) ion. If all the sodium, chloride, and calcium sulfate ($CaSO_4$) in sediment were dumped back into the ocean they would readily dissolve. However, salinities are considerably higher than normal in isolated arms of the ocean like the Persian Gulf, where water is evaporated more rapidly than it is added to the basin by rain or by runoff from adjacent lands and where mixing with the open ocean is sluggish. In some of these basins, the salinity rises to the point where $CaSO_4$ and even $NaCl$ precipitate. We know from the geologic record that most of the very large salt deposits found buried in sedimentary sequences were formed in a similar way (in arid peripheral basins which did not communicate readily with the sea and intermittently may even have been cut off entirely). Once such an evaporite deposit has formed, it can be preserved for hundreds of millions of years if it is buried under impermeable shale. Eventually, continental drift and the resultant crust upheavals bring these salt deposits back to the surface where fresh water can attack them. The contained salt is then dissolved and sent back to the ocean.

From this discussion we can see that the fractions of these two major anions that exist in the ocean at any one time depend on chance tectonic factors. How often is a large basin formed that has just the right geometry (and climate) to extract salt rapidly from the ocean? How often does a gigantic salt deposit thus formed come back into contact with fresh waters and dissolve? At times in the past when arid peripheral basins extracted salt at a higher than usual rate or when old deposits were being destroyed at a lower than usual rate, the salt content of the ocean must have decreased. When the reverse was true, the salt content must have increased. So the salt content of the ocean has probably fluctuated over the epochs of geologic time. Such fluctuations, however, have probably been fairly small because not enough salt is stored in the sediments to effect very large changes. For example, if all the NaCl and $CaSO_4$ now known to exist in sediments were dissolved in the sea, the salinity of the ocean would only increase by a bit more than a factor of 2.

Any fluctuations that have occurred must also have been very gradual. The cycling time for chloride and sulfur between sediments and the sea is very slow. From the rates at which these processes are currently taking place, we estimate that a Cl^- ion remains in the ocean for perhaps 200 million years before being "captured" in a basin where NaCl is actively precipitating. The Cl^- ion may reside in the sediment so formed for another 200 million years before returning to the ocean. Thus a typical Cl atom has spent about half its time in the sediments as NaCl and half its time in the ocean as a Cl^- ion. Sulfur, on the other hand, must spend a greater fraction of its time entrapped in sediments. A typical S atom remains in the ocean about 50 million years before it is fixed into $CaSO_4$ or iron sulfide (FeS_2). The S atom then remains in the sediments for about 200 million years before it is released by solution and carried back to the ocean.

Thus we see that the salinity of the ocean is fixed by two factors: the ratio of the elements hydrogen, chlorine, and sulfur in volatiles released from the earth's interior, and the tectonic factors influencing the partitioning of sulfur and chlorine between the ocean and the sediments (virtually all the water at the earth's surface is now, and more than likely always has been, in the ocean). The earth received these elements in a particular ratio from the solar nebula, and it is in roughly this ratio that they appeared at the surface. Since their arrival, the earth's geography has dictated their distribution between solid NaCl, solid $CaSO_4$, solid FeS_2, and dissolved sea salt.

Factors Controlling the Cation Content of the Sea

The ocean is electrically neutral. That means that once the anion content of the sea is fixed, the cation content is automatically determined. For each negative charge due to Cl^- and SO_4^{--} ions, a corresponding positive charge must exist. A great many cations could conceivably contribute to this balance. However, only four elements—sodium, potassium, calcium, and magnesium—are both sufficiently soluble in sea water and abundant in crustal material to make significant contributions. Just as SO_4^{--} and Cl^- contribute the major part of the negative charge, Na^+, K^+, Ca^{++}, and Mg^{++} contribute the major part of the positive charge.

Whereas the major anions in sea salt are found in only trace quantities in igneous rocks, the major cations are present in large quantities. Continental erosion therefore carries more K^+, Mg^{++}, Na^+, and Ca^{++} ions to the sea than it does SO_4^{--} and Cl^- ions. How then does the sea manage to maintain the required electrical balance? Somehow the ocean must dispense with the cation excess it continually receives from rivers by precipitating new minerals similar in chemical composition to those being weathered on the continents. If the amount of cations contained by the sea were to become a little too high (compared to the amounts of Cl^- and SO_4^{--} ions), the OH^- content (and hence the pH) of sea water would rise. This in turn would lead to a more rapid formation of the mineral phases that remove these cations. Some such self-balancing mechanism must exist to maintain equality between cations and anions within the sea. Unfortunately, we know almost nothing about the alumino-silicate mineral phases being formed in the sea. Until we do, arguments such as the above must remain quite vague.

The next question we might raise is what controls the proportions of these four cations in sea water. Let us begin by examining their relative abundance in river water and, in turn, in the rocks undergoing erosion. In river water (see Table 7-2), we find that for every mole of K there are 2 moles of Na, 1 mole of Mg, and 5 moles of Ca. In the ocean, for every mole of K there are 50 of Na, 5 of Mg, and only 1 of Ca. So rivers carry these ions to the ocean in much different proportions than are currently present in the sea. The reason the sea stores these cations in relative amounts which differ so radically from the amounts found in river water is that their removal tendencies from the sea are quite different. The sea rids itself of calcium most readily. This element is a constituent of the calcite produced in great quantities by marine organisms. After Ca, potassium removal is the most rapid. Apparently (we are not yet certain) K is taken up on the detrital alumino-silicate minerals brought to the sea by wind and water. By comparison, the sea has considerable difficulty getting rid of the sodium and magnesium it receives. The only way it can do so is to let such a very high content of Na and Mg build up relative to K and Ca that the minerals which form and the volcanic rocks which are altered on the sea floor are literally flooded with sodium and magnesium, carrying them away as fast as they enter the sea from the rivers.

The ratios of the four major cations in the sea thus depend upon

Table 7-2 Comparison between the chemical composition of sea water and the chemical composition of average river water. Corrections have been made for salt blown from the sea and for CO_2 transported via the atmosphere from the sea to soils.

Element	Mean Sea Water, moles/m³	Mean River Water, moles/m³	Sea / River
Na	500	.10*	5000
K	10	.05	200
Mg	50	.05	1000
Ca	10	.25	40
Cl	500	†	—
S	30	.06	500
C	2	.30‡	7
N	3×10^{-2}	—	—
P	2×10^{-3}	$\approx 4 \times 10^{-4}$	5
Si	.2	.04	5
U	1.5×10^{-5}	$.8 \times 10^{-6}$	20
Ba	1×10^{-4}	2×10^{-4}	.5

* Corrected for sea salt.
† Dominated by sea salt.
‡ Corrected for atmospheric CO_2.

the ratios in which these cations are being added from rivers and the difficulty the sea has in ridding itself of them. The former is fixed by the composition of continental rocks (and hence ultimately depends upon the volcanic liquids issuing forth from the earth's interior). The latter is fixed by the affinity of the minerals forming within the sea for a particular cation. For each unit of difficulty the ocean has in getting rid of Ca, the sea currently has 5 units of difficulty in ridding itself of K, 25 units of difficulty in ridding itself of Mg, and 125 units of difficulty in ridding itself of Na.

An analogy might help here. Four teams are competing in a game of musical chairs. The players walk around a circle until the music stops and then everyone dives for a chair. As there are many more participants than there are chairs, not every player gets a seat. The first team is comprised of professional athletes, the second of high-school students, the third of housewives, and the fourth of retired foundry workers. Clearly, in the first round, the athletes and the students will get the greater number of seats. Now, contrary to the normal rules in such a game, those who succeed in grabbing chairs step aside and let those who failed try again. In the second round, the housewives and foundry workers will fare considerably better against the diminished ranks of their opponents. Obviously, in our analogy, Ca and K are the

Table 7-3 Comparison of the cation composition of sea water with the cation composition of various inland seas. (Data compiled by H. J. Simpson, Lamont-Doherty Geological Observatory.)

Water Body	Na	K	Mg	Ca	Na/K	Mg/Ca	Na + K / Mg + Ca
			moles/m³				
Dead Sea (Israel)	1390	205	1150	350	7	3	1.1
Desmet (U.S.A.)	60	2.2	18	1.9	27	10	3.1
Bolac (Australia)	14	.2	3.3	.7	70	5	3.6
Caspian (USSR)	140	1.8	31	7.5	78	4	3.7
Tagar (USSR)	260	5.4	35	1.4	48	25	7.3
Timboram (Australia)	1570	10.2	175	34	154	5	7.6
Gnarport (Australia)	165	1.0	20	1.6	165	12	7.7
OCEAN	460	9.7	51	10	47	5	7.7
Murdeduke (Australia)	165	1.0	17	.6	165	28	9.4
Great Salt (U.S.A.)	3640	104	300	6.0	35	50	12
Corangamite (Australia)	300	2.6	25	.9	115	27	12
Pyramid (U.S.A.)	70	2.6	4.5	.2	27	22	15
Walker (U.S.A.)	100	2.4	4.3	.2	42	21	23
Bitter (USSR)	820	21	31	2.2	39	14	25
Little Borax (U.S.A.)	150	19	1.0	.2	8	5	33
Big Borax (U.S.A.)	270	7.5	1.0	.1	36	10	250
Albert (U.S.A.)	165	4.5	.5	.1	37	5	280
Mono (U.S.A.)	945	29	1.5	.1	33	15	610

athletes and the students of the ocean; Mg and Na are the housewives and retired foundry workers; and the sites in the lattices of forming minerals are the chairs. As the Ca and K ranks are depleted, the less reactive Mg and Na atoms find it easier to capture their share of lattice sites.

A comparison between the cation composition of the ocean and other saline water bodies in closed basins is revealing in this connection. As shown in Table 7-3, despite major differences in total salt content and anion composition, these saline lakes have Na/K and Mg/Ca ratios similar to those found in the ocean. The Na+K/Mg+Ca ratios, on the other hand, cover an enormous range, suggesting that the singly charged Na and K ions compete for a different set of mineral lattice sites than the doubly charged Mg and Ca ions do. In our analogy, it would be as if the businesspeople and the students were competing for chairs of one color and the housewives and the athletes for chairs of another color. The rivers feeding the saline lakes have chemical compositions reasonably similar to the average composition of those rivers feeding the sea. Any water body that must ultimately dispense the ions it receives to its sediments (rather than through an outlet to the sea) must have a Na/K ratio roughly 20 times and a Mg/Ca ratio roughly 50 times higher than the respective ratios in incoming river water. (In competing for sites accommodating singly charged ions, Na is only about one-twentieth as effective as K. In competing for doubly charged ion sites, Mg is only one-fiftieth as effective as Ca.)

Many attempts have been made to determine in what form Mg and Na leave the ocean, yet no universally accepted mode of extraction has been found for these two elements. This constitutes one of the major unanswered questions in chemical oceanography. If Mg and Na are being removed in the sediment, then either the host minerals are amorphous and therefore cannot be identified by x-rays or the host minerals are being mixed with so much detrital material that they are diluted beyond recognition. Another possibility is that these elements are fixed into alteration products of the basalts which constitute the initial lining of the sea floor. Since we know little of the extent (either laterally or vertically) of such alteration, there is no way to quantify the importance of this mechanism.

Still another possibility is that the chemical alteration of sediments long after their formation removes Mg and Na from the pore fluids surrounding the mineral grains and that these elements are continually resupplied to the depleted pore fluids by diffusion from the overlying sea. Considerable evidence has been obtained to show that pore water depletion does occur in the case of Mg. Measurements of the Mg content of pore fluids reveal a decrease with depth in the sediment column (see Figure 7-1). The missing Mg either reacts with $CaCO_3$ to form dolomite ($MgCa(CO_3)_2$), which appears as a secondary mineral at depth within most sediments, or in part replaces the iron which is removed from oxide or silicate minerals to form iron sulfide (a secondary

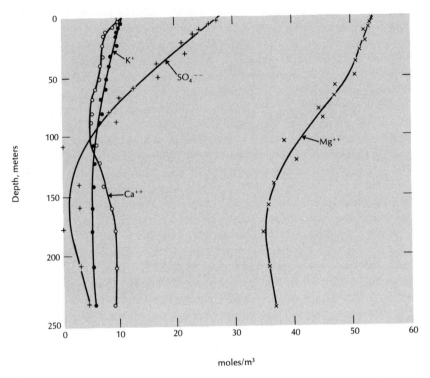

Figure 7-1   The K^+, Mg^{++}, Ca^{++}, and SO_4^{--} ion contents of pore water squeezed from various depths in a core taken at $13°24'$ N latitude and $63°45'$ W longitude and at a water depth of 1223 kilometers in the Caribbean Sea as part of the National Science Foundation's Deep Sea Drilling Program. (Data obtained by Fred Sayles and Frank Manheim, Woods Hole Oceanographic Institution.)

mineral resulting from the conversion of SO_4^{--} ion to S^{--} ion by anaerobic bacteria). Figure 7-2 shows the correlation between SO_4^{--} and cation loss in pore water samples from deep sea drill holes. But can this process account for the removal of a major fraction of the Mg and the Na entering the sea? We simply do not know.

So we can see that our knowledge of the cycles of the major cations is surprisingly primitive. Most of the hypotheses proposed rest on inadequate evidence.

Controls for the Limiting Nutrients

Next we will consider two of the biologically utilized elements, phosphorus and silicon, to determine what might control their concentrations in the sea. We have already noted that the phosphorus contained in

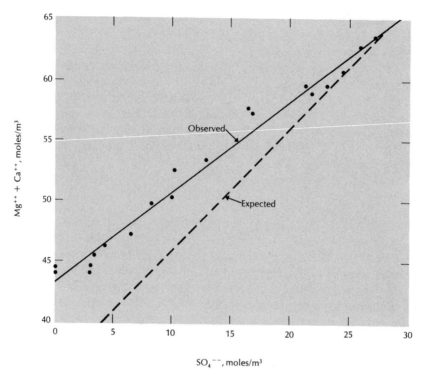

Figure 7-2 Plot of the Mg^{++} + Ca^{++} versus SO$_4^{--}$ ion contents of pore waters squeezed from sediments from the drilling site in Figure 7-1. The solid line represents the average relationship derived from these measurements. The dashed line represents the relationship expected if Ca^{++} and Mg^{++} were the only ions replacing the iron removed from oxide minerals for precipitation by the reduced sulfur. The shallower slope of the observed line indicates that ions other than Ca^{++} and Mg^{++} are involved. Potassium is one candidate, but the decrease in K$^+$ (as shown in Figure 7-1) accounts for only one-third of the slope difference. The logical candidate for the remaining difference is Na$^+$. Since the Na$^+$ concentration is so large, the loss of the needed amount would not produce a measurable gradient in the pore water column.

the sea is being renewed roughly once every 100,000 years; in other words, each year rivers add 1 part in 10^5 of the total amount of phosphorus contained by the sea. Our fundamental assumption has been that the amount of phosphorus leaving the sea must, at least over a period of several hundred thousand years, be equal to the amount added to the sea by rivers. Also, for every unit of P brought to the surface ocean by river water, about 100 units return from depth. The P arriving at the surface is almost entirely fixed into particles that fall to the deep sea. A small fraction of these particles survive destruction and are buried in the sediments.

With these facts in mind, we can see how the P content of the ocean is controlled: the abundance of P in the sea rises to a level where plants introduce just enough organic matter into the food chain that the fraction of this organic matter that escapes destruction carries exactly the amount of P added by the rivers! If some perturbation were to cause P to be removed from the sea faster than it is added, the P content of the sea would gradually fall. As it fell, plant productivity would decline, as would the amount of organic residue surviving destruction. The P content would continue to decrease until loss once again matched input.

Using the equations given in Chapter 1, we note that P balance is achieved when:

$$[P]_{deep} = \frac{v_{river}}{v_{mix}} \frac{1}{f} [P]_{river}$$

Hence, the amount of phosphorus dissolved in the sea is fixed by a combination of (1) the upwelling rate of deep sea water, (2) the fraction of particles falling to the deep sea that survive oxidation, (3) the P content of average river water, and (4) the rate of continental runoff. A change in any one of these will be reflected by a change in the steady-state P content of the sea.

The amount of rain that falls (and runs to the sea) and the rate at which the ocean waters mix both depend in a complex way on the earth's climate and on the arrangement of the continents. Thus we might expect the v_{river}/v_{mix} ratio to have changed as the climate changed and as the continents shifted their positions relative to one another and to the poles. If the v_{river}/v_{mix} ratio has changed, then the P content of the sea must also have changed.

The P content of river water depends on the composition of the rock being weathered and on the corrosive power of the surface waters. Because of changes in vegetation patterns, topographical relief, and surface rock composition, it is likely that, in the past, each unit of river water contained different amounts of phosphorus than it does today. If this is true, then the composition of sea water must have been correspondingly different.

The fraction of particles surviving destruction in deep water must depend on the water's oxygen content. Clearly, if no oxygen were present in the deep water, no animals or aerobic bacteria would exist to chew away at the falling organic matter. Instead the job would be done by the much less efficient anaerobic bacteria that use SO_4^{--} ion rather than O_2 gas as an oxidation agent. A much greater fraction of the particles would therefore survive and be buried in the sediment. In general, the lower the oxygen content of the deep water, the greater the survival probability for a given unit of organic tissue.

When the O_2 content of the atmosphere was smaller than it is today (before long ago plants built it up to its present level), the O_2

content in the deep sea was also smaller. The fraction f of organic material destroyed must have been correspondingly different. Such a change in f would cause a change in $[P]_{deep}$.

Thus, in the past, depending upon global geography, climate, atmospheric O_2 content, and the P content of river water, the ocean could well have had a different phosphorus content than it does today. Since P resides in the ocean for only 100,000 years, the reaction time of its oceanic concentration to environmental changes would be a fraction of a million years.

A very similar argument can be made for the element silicon, which is being removed from the water by diatoms and radiolarians. The Si content of sea water adjusts itself to a value at which the number of diatoms and radiolarians that form and survive destruction remove an amount of silica equal to the Si supplied to the ocean by rivers. If the supply and removal rates become unbalanced, the Si content will change until a balance between Si input and loss is reestablished.

Other geochemists contend that the silicon content of Pacific Deep Water has risen to a value representing equilibrium with the clay minerals in the underlying sediment, and this has been supported by laboratory experiments. If sea water containing either very little or very much silica remains in contact with various clay (alumino-silicate) minerals, a silicon concentration similar to that found in the deep Pacific is approached after several days *regardless* of the initial Si concentration. Hence it can be argued that if the Si content of sea water *in situ* were to depart from this value, silica would quickly be gained from, or lost to, the sediment until equilibrium was restored.

This hypothesis lost support when studies of the Dead Sea (a land-locked lake on the Israel–Jordan border, so salty that it cannot support diatoms) revealed that it has a higher SiO_2 content than the ocean and that its sediment pore waters have a much *lower* Si content than the overlying waters (see Table 7-4). Ocean sediment pore waters, by contrast, have a *higher* Si content than the overlying bottom water (the result of the gradual dissolution of diatoms and radiolarians

Table 7-4 Comparison between silica distribution in the waters of the Dead Sea and in the ocean. (Dead Sea data collected by D. J. Amit, Hebrew University, Israel.)

	Surface Water	Deep Water	Sediment Water*	Saturation Value†	
				Opal	Clay
	H_4SiO_4, moles/m^3				
Ocean	<.01	.15	.4	1.2	.4
Dead Sea	.25	.18	.05	2.0	.6

* Water squeezed from the intergranular pores.
† Higher for the 30° C warmer floor of the Dead Sea.

trapped in the sediment). In order for the Dead Sea to rid itself of the silica it receives, it has had to build up a Si content much greater than that necessary for equilibrium with its own sediment. In the ocean, the presence of silica-secreting organisms creates the opposite situation. While this observation makes it more likely that the Si content of the ocean is contingent upon the organic productivity of opal rather than upon the uptake and loss of silica by clay minerals, it is possible that both these mechanisms are at work. Part of the silica entering the ocean may be lost to minerals through inorganic uptake and part may be lost to opal through the actions of organisms.

Carbonate Ion Content Control

Now let us consider the situation for carbon. We must remember that carbon is leaving the sea in two quite different forms—as organic tissue and as $CaCO_3$. Thousands of analyses have been made of the relative amounts of these two forms of carbon found in ancient sediments. These studies reveal that, throughout the last several hundred million years, an average of roughly 20 percent of the carbon has entered the sediments in the form of organic material and 80 percent has entered the sediments in the form of $CaCO_3$. The amount of carbon entering the sediments in the form of organic tissue is very likely controlled by the cycle of phosphorus. For each unit of P removed to sediment a corresponding amount of C is lost. Since the C/P ratio in the rivers supplying the sea is considerably higher than the C/P ratio in the organic debris lost to the sediments, only part of the oceanic carbon can be removed in this way. The excess carbon (the total minus the carbon lost along with phosphorus as an organic tissue residue) must ultimately leave in the form of calcite precipitated by marine organisms.

We must then seek the mechanism which regulates the flow of $CaCO_3$ to the sediments. Organisms can produce as much $CaCO_3$ per unit of phosphorus as they wish because there is a superabundance of both calcium and carbon in the water. Today organisms are actually producing $CaCO_3$ about 6 times faster than rivers are making carbon available for this purpose. If all of this $CaCO_3$ were allowed to accumulate in the sediments, the ocean would run an enormous C deficit: it would be losing carbon several times faster than it receives this element. Somehow this process must be self-adjusting, and, in fact, it is. As we saw in Chapter 2, about four-fifths of all the $CaCO_3$ that precipitates reaches bottom in waters undersaturated with respect to $CaCO_3$ and largely redissolves. The remaining one-fifth falls onto areas along ridge crests and continental margins that project upward into the super-saturated water, and this $CaCO_3$ is largely preserved. If more $CaCO_3$ were preserved in the sediment than the amount required to balance the input of excess carbon, the ΣCO_2 in sea water would fall. This in turn would cause the CO_3^{--} ion content of the sea to fall.* The depth of the boundary between undersaturated and supersaturated water would rise. More area of the sea floor would then be covered by undersaturated water, a greater fraction of the $CaCO_3$ would be dissolved, and equilibrium would be reestablished.

Thus, if organisms were to start precipitating $CaCO_3$ tomorrow twice as fast as they are today (or 12 times faster than carbon is currently being made available for this purpose), the ratio of the area covered by red clay to the area covered by $CaCO_3$ ooze would gradually rise from 4 units of red clay per unit of $CaCO_3$ ooze to 7 units of red clay per unit of $CaCO_3$ ooze. This relationship is given in equation form as:

$$\frac{A_{CaCO_3}}{A_{CaCO_3} + A_{red\ clay}} = \frac{I_C - I_P \left(\dfrac{C}{P}\right)_{organic\ debris}}{P_{CaCO_3}}$$

where $A_{red\ clay}$ and A_{CaCO_3} are the areas covered by red clay and $CaCO_3$ ooze, I_C is the rate at which carbon is added to the sea, I_P is the rate at which phosphorus is added to the sea, C/P is the ratio of carbon to phosphorus atoms in the organic detritus in sediments, and P_{CaCO_3} is the production rate of calcite. This equation states that the fraction of the area of the deep sea floor covered by $CaCO_3$-rich mud must be equal to the ratio of the rate at which carbon is made available for removal as $CaCO_3$ (the total rate of carbon addition minus the rate of carbon destined to become organic matter) to the rate at which $CaCO_3$ is being manufactured by organisms.

So we see that there is a fundamental difference in the regulation of the P and the C cycles. The flow of phosphorus is regulated by controlling the rate at which organic tissue is formed. The flow of carbon is regulated by controlling the degree to which $CaCO_3$ is destroyed!

The Sedimentary Record

It seems logical to assume that changes in ocean chemistry are recorded in marine sediments. Although this is true, the record is far less evident than might be supposed. The reason is that the overall chemical composition of the sediments is not dependent on the chemical composition of the sea but on the composition of what is being added to the sea.

---

* The alkalinity would decrease twice as fast as the ΣCO_2 would. Since the CO_3^{--} ion content is equal to the difference between these two quantities, it would also have to decrease.

Since most of the matter added to the sea remains there a relatively short time, geologically speaking, the bulk composition of what leaves cannot be different from the bulk composition of what enters the sea. As we saw earlier, the composition of the sea adjusts itself to this input so that material is dispensed to the sediments at the same rate it enters the ocean. If we are to read the chemical record of the ocean from its sediments, our approach must therefore be more subtle than a bulk chemical analysis of the sediments of various ages.

We have already seen that the distribution in the sediment of various components produced within the sea depends on the interaction of the mode of mixing with the mode of production and destruction of organic debris. For example, the distribution of $CaCO_3$ reflects not only the fact that organisms produce more of this substance than rivers supply, but also the fact that in today's ocean the deep water flows from a surface source in the North Atlantic through the deep Indian Ocean and finally to the deep Pacific. The consequent Pacific-ward enrichment of the nutrient elements leads to a tilt in the level of the horizon separating $CaCO_3$ supersaturated and $CaCO_3$ undersaturated waters that preferentially deposits $CaCO_3$ in the Atlantic Ocean.

Thus, paleocalcium carbonate distribution maps would reveal changes in the mode of operation of the sea. If, for example, the fraction of the ocean floor covered with sediments rich in $CaCO_3$ were smaller at some time in the past, then we could surmise that the product of Ca^{++} and CO_3^{--} concentrations in the ocean was lower at that time than it is today. If the time were, let us say, 20,000 years ago during the height of the last ice age, we could attribute the change to a reduction in CO_3^{--} (the residence time of Ca in the sea is a million years, so its concentration could not change in such a short time). If the time in question were many millions of years ago, then either or both of these ions could have had different concentrations. The cause of such changes would be a difference in the ratio of the rate of $CaCO_3$ production by marine organisms to the rate at which the ocean is supplied with carbon from rivers. This could occur as a result of changes in the rate of oceanic mixing, in the amount of $CaCO_3$ manufactured per unit of limiting nutrient brought to the surface, or in the rate of continental erosion. Independent evidence would be required to distinguish among these possibilities.

If, on the other hand, at some time in the past the relative depth of $CaCO_3$ preservation was found to be greater than it is at present in the Pacific and shallower than it is at present in the Atlantic, then a change in the mode of circulation would be indicated. The implication would be that the main source of deep water was not in the northern regions of the Atlantic, as it is today. If this were the case, not only the distribution of $CaCO_3$ but also the distribution of SiO_2 in the sediments would have differed. Today the Atlantic is silica-starved and little opal is found in its sediment, but a more equitable distribution of

Table 7-5   Response times of various substances dissolved in the sea to changes in the factors determining their steady-state concentrations. Excepting CO_3^{--} ion and alkalinity, the response times are calculated by dividing the amount of the substance contained by the sea by the rate at which that substance is being added to (or removed from) the sea.

Substance	Response Time, Years
Na^+	200,000,000
Mg^{++}	40,000,000
SO_4^{--}	20,000,000
K^+	8,000,000
Ca^{++}	2,000,000
U	800,000
ΣCO_2	300,000
PO_4	200,000
H_4SiO_4	200,000
Alkalinity	100,000
CO_3^{--}	15,000

opaline-rich sediments between these two oceans might be expected if NADW were not pushing the silica toward the Pacific.

While the distribution of sediment types depends on the chemistry of the sea, we see that it does so in a complex manner. The dynamics of ocean mixing play an extremely important role. Changes in sediment distribution indicate that the system changed its mode of operation. These changes may even provide clues to the paleocirculation pattern and CO_3^{--} ion content. However, far more information is needed if we are to characterize the chemistry of the sea at some time in the past.

What other indicators are there? Only a very few have been found. One involves an element we have mentioned but whose chemistry we have not considered. Uranium is present in the sea in remarkably great quantities compared to its abundance in crustal rocks. It is a bio-unlimited element with a reactivity so low that it remains in the sea for about 800,000 years before being removed to the sediment (see Table 7-5).

Measurements on saline lakes reveal that they have high uranium contents. In fact, the U content of these lakes is proportional to the amount of total dissolved carbon they contain. Pyramid Lake in Nevada, for example, has a tenfold higher ΣCO_2 and a tenfold higher U content than sea water. Mono Lake in California has a hundredfold higher ΣCO_2 and a hundredfold higher U content. We can imply from this that if, in the past, the ocean had a different ΣCO_2, it very likely had a different U content then, too. Uranium owes its remarkably low reactivity in natural waters and its correlation with ΣCO_2 content to the fact that it forms a highly stable complex with CO_3^{--} ion.

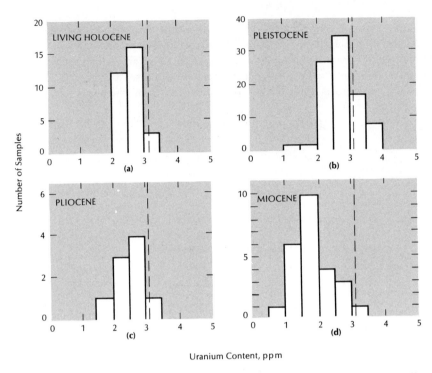

Figure 7-3 Histograms showing the uranium contents of corals in four different age groups. In each case, the U content expected if the corals formed with the same U/Ca ratio found in today's ocean is indicated by a dashed line. The data are shown in (a) for living corals and those formed during the last 8000 years (the present interglacial period); in (b) for corals which lived during the two previous periods of interglaciation (220,000–80,000 years in age); in (c) for corals from the Pliocene Epoch (2–5 million years in age); and in (d) for corals from the Miocene Epoch (5–20 million years in age).

Uranium has another peculiarity. Corals incorporate uranium in their $CaCO_3$ in nearly the same proportion to calcium as the U/Ca ratio found in sea water. Apparently, these organisms cannot distinguish between Ca, Sr, Ba, Ra, and U, and use them indiscriminately in building their $CaCO_3$ reefs. Thus the U/Ca ratio in a well-preserved marine coral should be an index of the U/Ca ratio in ancient sea water.

More importantly, since the U content of the sea apparently reflects its CO_3^{--} ion content, if corals formed with a different U/Ca ratio at some time in the past, then the sea in which they formed probably had a different CO_3^{--}/Ca^{++} ratio!

Figure 7-3 shows histograms of the U contents of corals for four time intervals: (a) the last 8000 years, (b) 80,000–220,000 years ago, (c) 2–5 million years ago, and (d) 5–20 million years ago. No significant differences are found back to 5 million years. The U/Ca ratios in corals from each interval in this time period are much the same as the U/Ca ratios in the living corals in today's ocean. In the 5–20 million year interval, however, the U/Ca ratio is an average of only half as great as it is in living corals. That the corals used for this study are well preserved (no conversion of unstable aragonite to calcite has occurred) and that they yield uranium–helium ages* consistent with those based on the absolute geologic ages of the deposits in which they are found suggest that the low U content of the Miocene corals is not the result of U loss during the millions of years which have elapsed since these corals formed.

From studies of marine sediment borings we know that during this 5–20 million year time period the saturation horizon separating $CaCO_3$-rich and $CaCO_3$-poor sediments was at approximately the same elevation as it is today. We can conclude then that the product of the Ca^{++} and CO_3^{--} ion contents was also roughly the same. If this is true, to have had a twofold lower CO_3^{--}/Ca^{++} ratio, the Ca^{++} ion content must have been 1.4 times as great as it is today and the CO_3^{--} ion content must have been .7 times as great.

This in turn suggests that the Ca content of sea water must have decreased during the last 20 million years. If the anion content $(Cl^- + SO_4^{--})$ has remained the same, then the Ca decrease must have been accompanied by an increase in one of the other cations. Magnesium is a likely candidate. Could it be that the reason no satisfactory sink for Mg has been found in present-day sediments is that it is indeed not leaving the sea nearly as fast as it arrives? Since the magnesium buildup would be compensated by a loss of calcium, perhaps our conclusion in Chapter 2 that Ca is leaving the sea at the same rate it arrives should be challenged. Could Ca actually be leaving somewhat faster than it is arriving? The hypothesis could be made that the greater expanse of shallow water seas which existed in the early Cenozoic Era enhanced the reactivity of Mg and that, as these environments were gradually reduced by tectonic movement and the growth of icecaps, the reactivity of Mg dropped. The ocean could be compensating for this change by gradually increasing its Mg/Ca ratio. An interesting speculation!

Perhaps the most definitive indicator we know of paleochanges in ocean chemistry is the C-13/C-12 ratio in marine shells. Measurements show that the ratio of these two stable isotopes of carbon in shells is very close to their ratio in the dissolved carbon of the water in which the organisms producing the shells live (unlike the O-18/O-16 ratio, which shows a large and temperature-dependent difference). In particular, the shells of organisms living in the surface Pacific and on the Pacific floor record a difference of 2 per mil in the C-13/C-12 ratios in surface and in deep water. As we will see, this 2 per mil differ-

---

* The uranium-helium method is closely analogous to the potassium-argon method we discussed in Chapter 3.

ence is fixed by the C/P ratio in sea water. Hence differences between the C-13/C-12 ratios for planktonic and benthonic* forams coexisting in ancient sediments provide a record of the C/P ratio in the sea water at the time they were alive.

Before we examine this record, let us consider the reason why the C-13/C-12 ratio is related to the C/P ratio in deep water. Surface water carbon has a higher C-13 content than deep water carbon because the carbon in the organic tissue falling from surface water is *depleted* by 20 per mil in C-13 with respect to the carbon in the water itself (C-12 atoms are fixed 1.020 times more rapidly during photosynthesis than C-13 atoms are). We have seen that in today's ocean roughly 10 percent of the carbon arriving at the surface is fixed into organic tissue and removed as particulate organic debris. If C-13 is to be conserved (as it must be), then the remaining 90 percent of the carbon must be 2 per mil *enriched* in C-13 (that is, it must contain the C-13 left behind by the organisms). As we saw in Chapter 1, the amount of carbon removed in organic tissue reflects the ratio of C/P in falling debris to C/P in deep sea water. In today's ocean, only 10 percent of the available carbon is needed in organic tissue to match the available phosphorus. The mathematical relationship between the per mil difference $\triangle$ in the C-13 contents of carbon from surface water and carbon from deep water (and hence between carbon from planktonic and benthic shells) is:

$$\triangle = \frac{(C/P)_{\text{organic debris}}}{(C/P)_{\text{deep water}}} \; 20 \text{ per mil}$$

where 20 per mil is the magnitude of the photosynthetic separation of the carbon isotopes. Direct evidence based on the difference between the C-13/C-12 ratios in coexisting $CaCO_3$ and organic tissue from ancient sediments suggests that this fractionation factor has not changed over the last several hundred million years. Hence we can assume that the fractionation factor was 20 per mil for previous geologic epochs as well as for the present epoch. If changes in the benthic-planktonic $\triangle$ are found in the sedimentary record, then either the C/P ratio in falling organic debris or more likely the C/P ratio in the sea as a whole must have been different than it is now.

Figure 7-4 shows C-13/C-12 ratio data for planktonic and benthic forams from Pacific Ocean sediments over the last 3 million years. The benthic ratio has remained nearly constant, suggesting that the isotopic composition of average marine carbon has not changed. The planktonic ratio has undergone both long (hundreds of thousands of years) and short (tens of thousands of years) changes. The $\triangle$ value between benthic and planktonic forams 2.5 million years ago was twice as great

---

* Planktonic forams live in the surface ocean; benthonic forams live on the ocean floor.

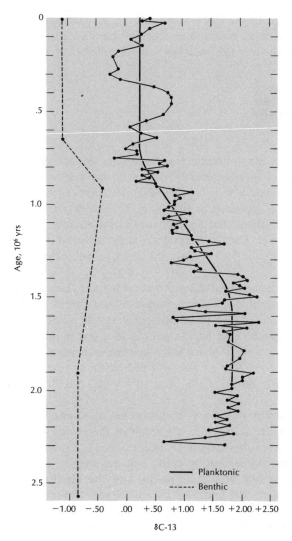

Figure 7-4 The C-13/C-12 ratio (expressed as δC-13) in planktonic and benthic foram shells from two deep cores covering an age range from the present back to 2.5 million years ago. The difference in the isotopic compositions of the shells of surface- and bottom-dwelling animals primarily reflects the limiting nutrient to carbon ratio in mean sea salt.

as it is today. If this reflects a change in the chemistry of the ocean, then the C/P ratio 2.5 million years ago must have been twice as great as it is today; that is, 20 percent rather than 10 percent of the carbon reaching the surface was fixed into falling organic debris at that time.

In summary, there are definite indications that the chemical composition of the ocean was not always the same as it is now. Differences in the distribution of $CaCO_3$ and SiO_2 in sediments, in the U content of corals, and in the C-13/C-12 ratio in planktonic shells provide evidence that at least the concentrations of the biologically active elements have changed over the past few tens of millions of years.

The Ocean's Future

The ocean is man's ultimate garbage can. Sooner or later, all of the products of civilization find their way to this reservoir. Will our waste disposal methods lead to any significant changes in ocean chemistry?

The best approach to this problem is to divide the pollutants into three groups: those that will add to the salt matrix (Cl^-, SO_4^{--}, Na^+, Mg^{++}, K^+, and Ca^{++}); those intimately involved with the life cycle (N, P, C, O_2, $\cdots$); and the trace constituents (heavy metals, organics, $\cdots$).

Even if man turned his entire effort toward changing the salt matrix of the sea, he could not. Since nature herself can dent this matrix only on a time scale of tens of millions of years, it would be presumptuous of man to assume that he can alter it in a few hundred years. All the $NaCl$, $CaSO_4$, $\cdots$ that we mine and use is being sent to the sea, but we will never be able to measure the resulting change in salinity. To date, we have mined about 4×10^{14} g of $NaCl$. The sea contains 2×10^{22} g!

Pollution by nutrient elements occurs in a variety of ways. Carbon and nitrogen are released as the result of the combustion of fossil fuels (and, of course, atmospheric O_2 is consumed). Exhausts from electrical power plants, home heating units, and automobile engines carry CO_2, CO, and a variety of nitrous oxides into the air. From the air, these substances eventually find their way to the sea. Nitrogen and phosphorus are the key ingredients in fertilizers. Just as in the ocean, the availability of N and P limits plant growth on the continents. To enhance plant productivity, farmers use copious amounts of these substances. Much of the farm produce goes to urban centers. From there, the N and P contained in the agricultural products are sent (as sewage) down rivers to the sea. Phosphate salts have traditionally been used as water softeners in detergents. This PO_4 also finds its way through sewage systems and rivers to the sea. What impact will the addition of all these nutrient elements (and the removal of O_2) have on the ocean–atmosphere system?

Oxygen is the element most easily evaluated. To date, the combustion of fossil fuels has consumed one-tenth of 1 percent of the O_2 contained in the atmosphere. When all the currently known chemical fuel reserves have been depleted, we will have consumed only about 2 percent of the available O_2. As this supply of fuel is enough for at least the next 100 years, there is certainly no imminent danger of making a sizable change in the atmosphere's O_2 content. Fortunately, ancient plant life provided us with more than enough O_2 to burn the chemical fuel available. Plenty of O_2 will be left for breathing!

An evaluation of CO_2 is less clear. The amount of CO_2 released by man's combustion of fossil fuels as of 1970 was equal to 20 percent of the amount currently in the atmosphere. When all the available fuel has been consumed, about 4 times the present amount of CO_2 will have been released in the atmosphere! How much of this CO_2 will enter the ocean? What will its effect be on marine chemistry? Measurements of the CO_2 content of the air over the island of Hawaii reveal that between 1958 and 1970 the annual increase has been only about 60 percent of that predicted if all the CO_2 produced had remained in the atmosphere (see Figure 7-5). There are three possible sinks for the "missing" CO_2: the ocean, the soils, and the living terrestrial biosphere. Of these, the ocean is by far the most important.

The ocean absorbs CO_2 from the atmosphere by the following reaction:

$$CO_2 + CO_3^{--} + H_2O \rightarrow 2\ HCO_3^{-}$$

Man's carbon dioxide gas combines with carbonate ion in the ocean to form bicarbonate ions. To see how this reaction alters the composition of surface water, we must consider the equilibrium requirement that:

$$k = \frac{[HCO_3^{-}]^2}{[CO_3^{--}]\,[CO_2]}$$

Because about 10 times more HCO_3^{-} ion than CO_3^{--} ion is present in the surface ocean to the first approximation, the CO_3^{--} ion content of surface ocean water will fall by the same fraction that its CO_2 content rises (that is, the product $CO_2 \times CO_3^{--}$ will remain nearly constant).

If 60 percent of all the CO_2 generated by man has remained in the atmosphere, then, as of 1970, the CO_2 content of the atmosphere (and of surface ocean water) was 12 percent higher than it was prior to the Industrial Revolution. Ocean water in contact with the air should then have had roughly a 12 percent lower CO_3^{--} ion content. In order to determine the amount of CO_2 required to reduce the CO_3^{--} ion content of the surface ocean by 12 percent, we must have a knowledge of the depths to which man's CO_2 has penetrated the ocean. The average "age" (or the time elapsed since release by production) of CO_2 molecules produced by man is about 25 years. In other words, the distribution of man-made CO_2 between the atmosphere and ocean is about the same as it would have been if all the CO_2 had been released in a single burst 25 years ago. From our knowledge of the vertical distribution of the bomb-produced isotopes (Sr-90, Cs-137, H-3, $\cdots$), we can estimate that the average depth of CO_2 penetration into the sea in 25 years

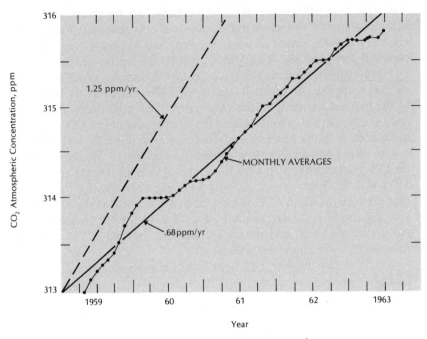

Figure 7-5 The CO_2 content of the atmosphere as observed at Mauna Loa Observatory on the Island of Hawaii. The plotted curve is a twelve-month running mean which removes the annual CO_2 cycle (due to seasonal uptake and release by plants and by the ocean.) The solid line shows that the average rate of increase has been .68 ppm/yr between 1958 and 1964. If all the CO_2 produced by man during this interval had remained in the atmosphere, the increase would have been 1.25 ppm/yr (the dashed line). (Data collected by J. C. Pales and C. D. Keeling, Scripps Institution of Oceanography.)

is about 400 meters. The amount of CO_3^{--} ion in an ocean layer this thick is about 80 moles/m². Thus a 12 percent rise in atmospheric CO_2 content would lead to about a 10 mole/m² reduction in CO_3^{--} ion content. Since one mole of CO_2 is removed from the atmosphere for each mole of CO_3^{--} ion that is destroyed, 10 moles of man-made CO_2 are removed by each square meter of sea surface. A 12 percent increase in atmospheric CO_2 amounts to 18 moles/m² of ocean surface. The fraction of CO_2 taken up by the ocean would therefore be 10/10 + 18, or 36 percent. Thus oceanic uptake can account for most of the CO_2 that does not appear in the atmosphere!

If man's demand for chemical energy continues to rise by about 5 percent per year, as it has in the last decade, then the mean age of man-made CO_2 will not change. The average time available for penetration into the ocean will remain close to 25 years. Hence the fraction of CO_2 entering the ocean should also remain roughly .36. In about

1990, when man has consumed enough fuel to produce an amount of CO_2 twice the amount produced by 1970, the atmosphere will have a 24 percent higher CO_2 content and the surface ocean a 24 percent lower CO_3^{--} ion content than before the Industrial Revolution.

Is this gradual drop in CO_3^{--} ion content detrimental to the ocean? It is hard to see how such a small change could greatly affect ocean life. The only conceivable consequence is that eventually the surface water will become undersaturated with respect to $CaCO_3$, and this may (although we are not yet certain) endanger some of the animals and plants that form calcitic and aragonitic cages. We have seen that the surface ocean today is about sevenfold supersaturated with respect to aragonite. Even when man has burned up all the known chemical fuel reserves, the saturation point will probably not have been passed. In any case, in the interim, we will have adequate time to determine if various marine organisms will be adversely affected. In all likelihood, the climatic changes produced by the 60 percent of the CO_2 which remains in the atmosphere will present problems of much greater import, and decisions with regard to stemming the flow of man-made CO_2 will be based on the climatic changes induced by the atmospheric component rather than on the ecological changes induced by the oceanic component.

Man mines about 1×10^{13} g/yr of phosphate. He forms about 3×10^{13} g/yr of nitrate from atmospheric nitrogen. An additional 3×10^{13} g/yr of nitrate are inadvertently formed during the combustion of fuels. Vertical mixing in the sea brings 2×10^{14} g/yr of phosphate and 2×10^{15} g/yr of nitrate to the surface of the sea. Thus, even adding the yearly productions of all of these substances to the sea would have little impact on total plant production. However, since nitrate and phosphate reach the sea largely via rivers and estuaries, the transient effects in these small systems will, of course, be very large. Decisions regarding nutrient control will thus stem from the effects of nutrients on inland and coastal waters, not their effect on the open ocean.

Thus we see that, once mixed throughout the surface ocean, man's C, N, and P will not seriously alter ocean chemistry. Does this mean that if a method could be found to bypass inland and coastal waters, man could look to the sea as a reservoir for all his pollutants? Certainly not. First of all, anything added to the sea can no longer be used by man. Thus, by dumping valuable resources into the sea, we are squandering our natural heritage. Even more important in this respect is the third category of pollutants—trace metals and organics. Many of these substances (DDT, for instance) were absent from the sea until man began to manufacture them. Others (like mercury) are present in such small quantities that man is capable of overwhelming their natural levels. When we dump the byproducts of our civilization into the earth's air and waters, we are not only sending innocuous substances like chlorides and phosphates to the sea. A whole host of other substances are also being discharged. The effects of most of these accompanying substances on marine life and on the people who eat marine foods are not really known, and the manner in which these substances are dispersed, degraded, and removed to the sediments is also largely unknown.

The unknown—our ignorance—must be stressed. Our understanding of marine chemistry is limited largely to the most abundant constituents of sea salt and of organisms—perhaps the most interesting factors from a historical point of view, but the least important when the future is considered. Man's eventual impact on the sea will be made by highly reactive substances with concentrations so low and chemical pathways so complex that they have thus far defied analysis. If this book is rewritten a decade from now, it may deal much more with the marine chemistry of man's poisons and much less with the classical components of sea salt we have covered here.

SUPPLEMENTARY READINGS

Rubey, W. W. "Geologic History of Sea Water." *Bull. Geol. Soc. Amer.* 62 (1951):1111–47. A classic paper presenting boundary conditions on the manner in which the oceans originated and developed through time.

Sillen, L. G. "The Physical Chemistry of Sea Water." *Oceanography*, ed. M. Sears. Washington, D.C.: American Association for the Advancement of Science, 1961, Publication 67, pp. 549–81. This paper presents a model based on thermodynamic controls which has had a profound impact on the thinking in this subject.

Broecker, Wallace S. "A Kinetic Model for the Chemical Composition of Sea Water." *Quaternary Research* 1 (1971):188–207. Posits a theory of ocean chemistry controls based on kinetics rather than on strict thermodynamic equilibrium.

Two classic papers on ocean chemistry theory written by an Irish medical doctor prior to World War II.

Conway, E. J. "Mean Geochemical Data in Relation to Oceanic Evolution." *Proc. Roy. Irish Acad.* B48 (1942):119–59.

Conway, E. J. "The Chemical Evolution of the Ocean." *Proc. Roy. Irish Acad.* B48 (1943):161–212.

■ Publications concerned with the geochemistry of some properties used in the reconstruction of the chemical history of the oceans:

Kroopnick, P., Deuser, W. G., and Craig, H. "Carbon-13 Measurements on Dissolved Inorganic Carbon at the North Pacific (1969) GEOSECS Station." *J. Geophys. Res.* 75 (1970):7668–71. Discusses measurements of the vertical distribution of C-13/C-12 ratios in the ΣCO_2 of the Pacific Ocean.

Turekian, K. K., and Chan, L. H. "Marine Geochemistry of the Uranium Isotopes Th-230 and Pa-231." *Activation Analysis in Geochemistry and Cosmochemistry*, eds. Arild O. Brunfelt and Eiliv Steinnes. Universitetsforlaget, Oslo: Bergen and Tromsö, 1971, pp. 311–20. Deals with factors controlling the distribution of uranium and its isotopes in the sea.

Geochemical Investigations in the Caribbean Sea, Initial Reports of the Deep Sea Drilling Project, Volume 20. National Science Foundation, National Ocean Sediment Coring Program. Washington, D.C.: U.S. Government Printing Office, 1973. Gives the detailed chemistry of pore fluids extracted from three borings in the Caribbean Sea.

■ Papers dealing with the temporal history of the sea as reconstructed from studies of marine sediments:

Lowenstam, Heinz A. "Mineralogy, O-18/O-16 Ratios, and Strontium and Magnesium Contents of Recent and Fossil Brachiopods and Their Bearing on the History of the Oceans." *J. Geology* 69 (1961):241–60. An attempt to reconstruct the Sr, Mg, and O-18 concentrations in the sea from measurements made on fossil mollusks.

Ramsay, A. P. S. "Distribution of $CaCO_3$ in Deep Sea Sediments. *Studies in Paleo-Oceanography*, ed. W. W. Hay. Society of Economic Paleontologists and Mineralogists Special Publication. SEPM Research Symposium, 1971. A reconstruction of the variations in the level of $CaCO_3$ compensation over the past few tens of millions of years.

Berkner, L. V., and Marshall, L. C. "The History of the Growth of Oxygen in the Earth's Atmosphere." *The Origin and Evolution of Atmospheres and Oceans*, eds. P. J. Brancazio and A. G. W. Cameron. New York: John Wiley & Sons, Inc., 1964, pp. 102–26. An attempt to relate the evolution of marine and terrestrial organisms to the evolution of atmospheric O_2.

Broecker, Wallace S. "A Boundary Condition on the Evolution of Atmospheric Oxygen." *J. Geophys. Res.* 75 (1970):3553–57. An attempt to reconstruct the evolution of atmospheric O_2 from measurements of the C-13/C-12 ratio in fossil $CaCO_3$.

PROBLEMS

7-1 The fossil fuel CO_2 released by man is likely to increase the rate of continental weathering. If it does, the flow of cations to the sea will correspondingly increase. In what way will the sea respond to restore the balance between cation gain and loss?

7-2 If an ingenious method were developed which enabled man to increase the rate of vertical mixing throughout the sea by a factor of 2, what would be the immediate changes in:

(a) plant productivity?

(b) fish productivity?

(c) the rate of loss of phosphorus from the sea?

How would the balance between P loss and gain be restored? Assuming this twofold higher mixing rate continued indefinitely, over what time interval would the P balance be restored?

7-3   A fossil coral yields a uranium content twice that found in coral growing today. If, at the time it was created, the ancient coral was typical of the ocean and if the U content of that coral has not changed since it began to grow, then what were:

(a) the U/Ca ratio

(b) the CO_3^{--}/Ca^{++} ion ratio

(c) the Ca content

of the sea when the coral was created?

7-4   Fossil benthonic and planktonic forams are recovered from the 20 million year horizon in a deep sea core. The δC-13 value for the benthics is .0 per mil and, for the planktonics, 4.0 per mil. If the photosynthetic fractionation factor and the C/P ratio in marine organisms were the same 20 million years ago as they are now, by how much must the C/P ratio in the ancient deep sea have differed from the C/P ratio in today's sea?

7-5   Eight pounds of P are added to each acre of typical farmland every year. How many pounds of P would have to be added to each acre of sea surface every year to raise the supply by a factor of 10 over that derived from upwelling? What difficulties would be encountered in "farming" the sea that are not encountered in farming the land? How could large-scale aquaculture be carried out in the sea?

The drilling ship *Glomar Challenger*, operated by the Global Marine Corporation. An electronic beacon dropped to the bottom is tracked by a computer. Through the use of thrusters located in the bow and stern of the ship, the computer maintains the location of the *Challenger* to within 100 feet of a point directly above the beacon. This permits drilling without danger of snapping the pipe. Photograph courtesy of the Deep Sea Drilling Project, Scripps Institution of Oceanography.

CREDITS AND ACKNOWLEDGMENTS

p. 10 Photographs courtesy of (a) M. Roche, (b) T. Saito, (c) L. H. Burckle, and (d) P. H. Chen—all of Lamont-Doherty Geological Observatory.

p. 23 From A. C. Redfield, "The Biological Control of Chemical Factors in the Environment," *American Scientist* (1958), Vol. 46. Reprinted by permission *American Scientist*, Journal of Sigma Xi, the Scientific Research Society of North American, Inc.

p. 48 From Yuan-Hui Li, Taro Takahashi, and Wallace S. Broecker, "Degree of Saturation of $CaCO_3$ in the Oceans," *Journal of Geophysical Research* (1969), Vol. 74, No. 23, Figs. 9 and 10, p. 5521. Copyright by American Geophysical Union.

p. 66 From Wallace S. Broecker and Yuan-Hui Li, "Interchange of Water between the Major Oceans," *Journal of Geophysical Research* (1970), Vol. 75, No. 18, Fig. 4, p. 3548. Copyright by American Geophysical Union.

p. 73 From Wallace S. Broecker and Jan van Donk, "Insolation Changes, Ice Volumes, and the O-18 Record in Deep-Sea Cores," *Reviews of Geophysics and Space Physics* (1970), Vol. 8, No. 1, Fig. 2, p. 172. Copyright by American Geophysical Union.

207

pp. 76–77 From Charles Giffin, Aaron Kaufman, and Wallace S. Broecker, "Delayed Coincidence Counter for the Assay of Actinon and Thoron," *Journal of Geophysical Research* (1963), Vol. 68, No. 6, Fig. 1, p. 1750. Copyright by American Geophysical Union.

p. 81 From Wallace S. Broecker and Jan van Donk, "Insolation Changes, Ice Volumes, and the O-18 Record in Deep-Sea Cores," *Reviews of Geophysics and Space Physics* (1970), Vol. 8, No. 1, Figs. A and B, p. 176. Copyright by American Geophysical Union.

p. 95 From Wallace S. Broecker, A. Kaufman, and R. M. Trier, "The Residence Time of Thorium in Surface Sea Water and Its Implications Regarding the Rate of Reactive Pollutants," *Earth and Planetary Science Letters* (1973), Vol. 20, No. 1, p. 41.

p. 102 From M. L. Bender, Teh-Lung Ku, and Wallace S. Broecker, "Accumulation Rates of Manganese in Pelagic Sediments and Nodules," *Earth and Planetary Science Letters* (1970), Vol. 8, Fig. 1, p. 146.

p. 104 From Teh-Lung Ku and Wallace S. Broecker, "Uranium, Thorium, and Protactinium in a Manganese Nodule," *Earth and Planetary Science Letters* (1967), Vol. 2, Fig. 2, p. 319.

p. 106 From Teh-Lung Ku and Wallace S. Broecker, "Uranium, Thorium, and Protactinium in a Manganese Nodule," *Earth and Planetary Science Letters* (1967), Vol. 2, Fig. 3, p. 320.

p. 107 From Bruce C. Heezen and Charles D. Hollister, *The Face of the Deep*, p. 425. Copyright © 1971 by Oxford University Press, Inc. Reprinted by permission. Photograph courtesy of Charles D. Hollister, Woods Hole Oceanographic Institution.

p. 108 Photograph courtesy of R. K. Sorem and Ellen R. Foster, both of Washington State University.

p. 109 Photograph courtesy of Charles D. Hollister, Woods Hole Oceanographic Institution. Reprinted by permission of Polar Information Service, Office of Polar Programs, National Science Foundation, Washington, D.C.

p. 137 Charles Culberson, "Processes Effecting the Oceanic Distribution of CO₂," Ph.D. thesis, Oregon State University, June 1972.

p. 146 From Wallace S. Broecker and Virginia M. Oversby, *Chemical Equilibria in the Earth*, International Series in the Earth and Planetary Sciences, McGraw-Hill (1971), p. 167.

p. 147 From Wallace S. Broecker and Virginia M. Oversby, *Chemical Equilibria in the Earth*, International Series in the Earth and Planetary Sciences, McGraw-Hill (1971), p. 168.

p. 159 From H. Craig and R. F. Weiss, "The GEOSECS 1969 Intercalibration Station: Introduction, Hydrographic Features, and Total CO₂-O₂ Relationships," *Journal of Geophysical Research* (1970), Vol. 75, No. 36, Fig. 1, p. 7643. Copyright by American Geophysical Union.

p. 161 From H. Craig and R. F. Weiss, "The GEOSECS 1969 Intercalibration Station: Introduction, Hydrographic Features, and Total CO₂-O₂ Relationships," *Journal of Geophysical Research* (1970), Vol. 75, No. 36, Fig. 2, p. 7644. Copyright by American Geophysical Union.

p. 172 From Wallace S. Broecker and Tsung-Hung Peng, "Gas Exchange Rates between Air and Sea, *Tellus*, 1974 (in press).

p. 173 From A. Kaufman, R. M. Trier, Wallace S. Broecker, and H. W. Feely, "The Distribution of Ra-228 in the World Ocean," *Journal of Geophysical Research* (1973), Vol. 78, No. 36, p. 8832. Copyright by American Geophysical Union.

p. 174 From A. Kaufman, R. M. Trier, Wallace S. Broecker, and H. W. Feely, "The Distribution of Ra-228 in the World Ocean," *Journal of Geophysical Research* (1973), Vol. 78, No. 36, p. 8845. Copyright by American Geophysical Union.

p. 196 From Wallace S. Broecker, "A Kinetic Model for the Chemical Composition of Sea Water," *Quaternary Research* (1971), Vol. 1, No. 2, Fig. 4, p. 202. Copyright © 1971 by Academic Press, Inc.

p. 199 From Wallace S. Broecker, "A Kinetic Model for the Chemical Composition of Sea Water," *Quaternary Research* (1971), Vol. 1, No. 2, Fig. 5, p. 204. Copyright © 1971 by Academic Press, Inc.

p. 202 From J. C. Pales and C. D. Keeling, "The Concentration of Atmospheric Carbon Dioxide in Hawaii," *Journal of Geophysical Research* (1965), Vol. 70, No. 24, Fig. 11, p. 6066. Copyright by American Geophysical Union. Also appeared in Wallace S. Broecker, Yuan-Hui Li, and Tsung-Hung Peng, "Carbon Dioxide—Man's Unseen Artifact," in *Impingement of Man on the Oceans*, ed. D. W. Hood, Wiley (1971), p. 309.

Chapter Opening Photos

Chapter 1 Rosette sampler being lowered into the sea to collect 30-liter water samples for chemical analysis. The package is supported by a conducting cable that permits pressure, temperature, conductivity, oxygen, and light-scattering data from sensors contained within the ring of sampling bottles to be continuously telemetered to the deck of the ship. The samples are triggered at the desired depth by sending electrical pulses down the cable. The device shown here was used by the Geochemical Ocean Section Study (GEOSECS) in its global survey. The photograph was taken by Robert M. Trier, Lamont-Doherty Geological Observatory.

Chapter 2 Ocean bottom photograph taken with a camera lowered by cable to just above the sea floor. The sea spider is 27 inches in width. Trails of worms and of other benthonic scavengers can be seen etched into the red clay. This photograph from Woods Hole Oceanographic Institution.

Chapter 3 Piston coring device used to obtain sea-floor sediment sections up to 80 feet in length. The device is lowered by cable to about 50 feet above the sea floor. At this point, a bottom-feeling trigger allows the corer to free fall. In addition to the thrust provided by the two ton weight at the far end of the pipe, suction provided by a piston which stops just above the sediment–water interface aids in driving the sediment into the pipe. The corer shown here is one used by the Lamont-Doherty Geological Observatory. The photograph was taken by Robert M. Trier, Lamont-Doherty Geological Observatory.

Chapter 4 Chain link dredge used to obtain manganese nodules and outcropping rock from the deep sea. The soft sediment scooped up by the rectangular frame flows through the links. In this way, large objects are captured. The device shown here was designed by the Scripps Institution of Oceanography. The photograph was taken by Robert M. Trier, Lamont-Doherty Geological Observatory.

Chapter 5 Gas exchange between the ocean and atmosphere is promoted by rough seas. Spray enters the air—bubbles are entrained in the water. The scene shown here is an all too familiar one to the sea-going oceanographer. The photograph was taken by Corsini for Standard Oil Co. (N.J.).

Chapter 6 Large volume water sampler used to collect samples for C-14 measurement. A large circular door in the top of the sampler is held in the open position as the sampler is lowered. The hood scoops water into the sampler and causes it to be rapidly flushed. When it reaches the desired depth, a "messenger" weight drops down the wire, hits a trigger arm, and allows the spring-loaded door to snap shut. This sampler was designed at the Lamont-Doherty Geological Observatory. The photograph was taken by Robert M. Trier, Lamont-Doherty Geological Observatory.

Chapter 7 The drilling platform of the research vessel *Glomar Challenger*. To date, this ship has drilled through the sedimentary column at more than 300 locations on the sea floor as part of a program sponsored by the National Science Foundation. The pipe sections stored in the forward racks are assembled into a string of pipe that dangles several miles beneath the *Challenger* before reaching the sediment. Once there, the pipe can drill and recover sediment sections to depths exceeding 1000 feet beneath the sediment–water interface. A sedimentary column of this thickness can be drilled and sampled in one or two days. Photograph courtesy of the Deep Sea Drilling Project, Scripps Institution of Oceanography.

INDEX

A 4
B 5
C 6
D 7
E 8
F 9
G 0
H 1
I 2
J 3